PRAISE f

THE SECOND KIND OF IMPOSSIBLE

"A rare and compulsively readable blend of science and thriller, *The Second Kind of Impossible* tells of the quest to find a new type of matter that would rewrite the rules of reality. Paul Steinhardt, one of the world's leading theoretical physicists, takes readers on a wondrous odyssey across multiple decades and continents as, against all odds, he helps to topple scientific orthodoxy."

—Brian Greene, author of *The Elegant Universe*

"Scientists, smugglers, and spies—this book is an exciting and enlightening scientific detective story. The tale is about far more than a new form of matter; it is also a thrilling and wonderfully written look at how science works."

—Walter Isaacson, author of *Einstein*

"An epic account of two scientific triumphs: a thirty-year theoretical search for understanding and a real-world expedition into the wilds of Kamchatka. It is as if *The Origin of Species* and *The Voyage of the Beagle* had been published together in one volume."

—Freeman Dyson, author of *Maker of Patterns*

"A truly amazing adventure story, full of twists and turns, right up to the very end. It has my strongest recommendation."

—Sir Roger Penrose, author of *The Emperor's New Mind*

"An intriguing blend of science and international adventure. [Steinhardt] takes readers on a wild ride in search of a new kind of matter . . . full of intrigue and adventure, culminating with the epic Kamchatka journey. . . . A general audience can and should enjoy this original, suspenseful true-life thriller of science investigation and discovery."

—*Publishers Weekly*

"A gripping scientific quest . . . an admirable popular account of the quasi-crystal, an oddball arrangement of atoms that seems to contradict scientific laws . . . Steinhardt [is] a pioneer in the field and a fine writer."

—*Kirkus Reviews*

"[A] memoir and rollercoaster adventure, packed with discovery, disappointment, exhilaration and persistence. . . . This book is a front-row seat to history as it is made."

—*Nature*

"Part physics primer, part fantastic adventure . . . Steinhardt's affection and admiration for the journey's colorful cast of characters infuse every page. Although his excitement is palpable, he is also careful and methodical, often reminding himself that he could be wrong. *The Second Kind of Impossible* shows the benefit of a slow and steady approach to science, where determination and luck are just as important as insight."

—*Science News*

"A thrilling mix of scientific memoir and true detective story. Most importantly, it is a tale of the excitement that drove the author to extraordinary insights far outside his original area of expertise. . . . I refrain from recounting the many astonishing turns of events the team encountered. But suffice to say that a fiction writer could hardly have thought of better plot twists."

—*Physics Today*

"Cutting-edge science as high adventure."

—*Booklist*

"Steinhardt does a masterful job of making a complex subject more accessible. . . . The quest-filled narrative along with the author's casual style create an extremely readable work that gives insight into the work involved in scientific discovery."

—*Library Journal*

"*The Second Kind of Impossible* is a must-read. Even if you have no interest in quasicrystals or five-fold crystalline structures, Steinhardt's book is a delight. The 364-page book reads like a novel—and a fast-paced, well-written one at that. Steinhardt manages to maintain a quick and thrilling pace without skimping on the science behind the story."

—*Crystallography Times*

ALSO BY PAUL J. STEINHARDT

Endless Universe: Beyond the Big Bang
(with Neil Turok)

THE SECOND KIND OF IMPOSSIBLE

THE EXTRAORDINARY QUEST
FOR A NEW FORM OF MATTER

PAUL J. STEINHARDT

SIMON & SCHUSTER PAPERBACKS

NEW YORK LONDON TORONTO SYDNEY NEW DELHI

Simon & Schuster Paperbacks
An Imprint of Simon & Schuster, Inc.
1230 Avenue of the Americas
New York, NY 10020

First Simon & Schuster trade paperback edition January 2020

SIMON & SCHUSTER PAPERBACKS and colophon are
registered trademarks of Simon & Schuster, Inc.

For information about special discounts for bulk purchases,
please contact Simon & Schuster Special Sales
at 1-866-506-1949 or business@simonandschuster.com.

The Simon & Schuster Speakers Bureau can bring authors to your live event.
For more information or to book an event contact the
Simon & Schuster Speakers Bureau at 1-866-248-3049
or visit our website at www.simonspeakers.com.

Interior design by Ruth Lee-Mui
Maps by Paul J. Pugliese

Manufactured in the United States of America

7   9   10   8

Library of Congress Cataloging-in-Publication Data is available.

ISBN 978-1-4767-2992-3
ISBN 978-1-4767-2993-0 (pbk)
ISBN 978-1-4767-2994-7 (ebook)

To the curious and fearless
who defy convention
risking ridicule and failure
to pursue their dreams of discovery

CONTENTS

PREFACE

MIDDLE OF NOWHERE, NORTH OF KAMCHATKA, JULY 22, 2011:
I held my breath as the blue behemoth lurched its way down the steep
incline. It was my first day in the mad contraption, a weird-looking
vehicle with what looked like a Russian army tank on bottom and a
beat-up moving van on top.

To my amazement, our driver, Viktor, managed to make it all
the way down the hill without toppling over. He hit the brakes, and
our truck shuddered and shook to a halt at the edge of a riverbed. He
turned off the ignition, and muttered a few words in Russian.

"Viktor says this is a good place to stop," our translator announced.

I peered out the front window, but could not for the life of me see
what was so good about it.

Climbing out of the cab, I stood atop the enormous tank treads
to get a better view. It was a cool summer evening, approaching midnight. But it was still light out, a reminder of how far I was from
home. The summer sky never gets very dark so close to the Arctic
Circle. The earthy, pungent smell of decaying vegetation filled the air,
the unmistakable smell of the tundra.

I jumped off the tank treads into the thick, spongy muck to

stretch my legs when, suddenly, I was attacked from all sides. Millions and millions of ravenous mosquitoes were springing up from the muck, drawn to the carbon dioxide I was exhaling. I swiped frantically with my arms and turned this way and that to escape them. Nothing helped. I had been warned about the tundra and its perils. Bears, insect swarms, unpredictable storms, endless miles of muddy swells and ruts. But these weren't just stories anymore. This had become all too real.

My critics were right, I realized. I had no business leading this expedition. I was neither a geologist nor an outdoorsman. I was a theoretical physicist who belonged back home in Princeton. I should be working on calculations, with notebook in hand, not trying to lead a team of Russian, Italian, and American scientists on what was probably a hopeless quest in search of a rare mineral that had traveled billions of years through space.

How could this have happened? I asked, as I struggled against the ever-growing swarm. Unfortunately, I knew the answer: The crazy expedition had been my idea, the fulfillment of a scientific fantasy that had been occupying my mind for nearly three decades. The seed was planted in the early 1980s when my student and I developed a theory showing how to create novel forms of matter long thought to be "impossible," atomic formations explicitly forbidden by venerable scientific principles.

I had learned early on to pay close attention whenever an idea is dismissed as "impossible." Most of the time, scientists are referring to something that is truly out of the question, like violating the conservation of energy or creating a perpetual motion machine. It never makes sense to pursue those kinds of ideas. But sometimes, an idea is judged to be "impossible" based on assumptions that could be violated under certain circumstances that have never been considered before. I call that the second kind of impossible.

If one can expose the underlying assumptions and find a long-overlooked loophole, the second kind of impossible is a potential gold mine that can offer a scientist the rare opportunity, perhaps a once-in-a-lifetime opportunity, to make a transformational discovery.

In the early 1980s, my student and I discovered a scientific loophole in one of the most well-established laws of science and, exploiting that, realized it was possible to create new forms of matter. In a remarkable coincidence, just as our theory was being developed, an example of the material was accidentally discovered in a nearby laboratory. And soon, a new field of science was born.

But there was one question that kept bothering me: *Why hadn't this discovery been made long ago?* Surely nature had made these forms of matter thousands, or millions, or perhaps even billions of years before we had dreamed them up. I could not stop myself from wondering where the natural versions of our material were being hidden and what secrets they might hold.

I did not realize at the time that this question would lead me down the road to Kamchatka, an almost thirty-year-long detective story with a dizzying array of improbable twists and turns along the way. So many seemingly insurmountable barriers had to be conquered that it sometimes felt like an unseen force was guiding me and my team step by step toward this exotic land. Our entire investigation had been so . . . impossible.

Now we were in the middle of nowhere, with everything we had achieved up to this point at risk. Success would depend on whether we were lucky enough and skillful enough to conquer all of the unexpected obstacles, some of them terrifying, that we were about to confront.

PART I

MAKING THE IMPOSSIBLE POSSIBLE

ONE

IMPOSSIBLE!

PASADENA, CALIFORNIA, 1985: *Impossible!*

The word resonated throughout the large lecture hall. I had just finished describing a revolutionary concept for a new type of matter that my graduate student, Dov Levine, and I had invented.

The Caltech lecture room was packed with scientists from every discipline across campus. The discussion had gone remarkably well. But just as the last of the crowd was filing out, there arose a familiar, booming voice and that word: *"Impossible!"*

I could have recognized that distinctive, gravelly voice with the unmistakable New York accent with my eyes closed. Standing before me was my scientific idol, the legendary physicist Richard Feynman, with his shock of graying, shoulder-length hair, wearing his characteristic white shirt, along with a disarming, devilish smile.

Feynman had won a Nobel Prize for his groundbreaking work developing the first quantum theory of electromagnetism. Within the scientific community, he was already considered one of the greatest theoretical physicists of the twentieth century. He would eventually achieve iconic status with the general public, as well, because of his pivotal role identifying the cause of the *Challenger* space shuttle disaster and his two bestselling books *"Surely You're*

Joking, Mr. Feynman!" and *"What Do You Care What Other People Think?"*

He had a wonderfully playful sense of humor, and was notorious for his elaborate practical jokes. But when it came to science, Feynman was always uncompromisingly honest and brutally critical, which made him an especially terrifying presence during scientific seminars. One could anticipate that he would interrupt and publicly challenge a speaker the moment he heard something that was, in his mind, imprecise or inaccurate.

So I had been keenly aware of Feynman's presence when he entered the lecture hall just before my presentation began and took his usual seat in the front row. I kept a careful watch on him out of the corner of my eye throughout the presentation, awaiting any potential outburst. But Feynman never interrupted and never raised an objection.

The fact that he came forward to confront me after the talk was something that probably would have petrified many scientists. But this was not our first encounter. I had been lucky enough to work closely with Feynman when I was an undergraduate at Caltech about a decade earlier and had nothing but admiration and affection for him. Feynman changed my life through his writings, lectures, and personal mentoring.

When I first arrived on campus as a freshman in 1970, my intention was to major in biology or mathematics. I had never been particularly interested in physics in high school. But I knew that every Caltech undergraduate was required to take two years of the subject.

I quickly discovered that freshman physics was wickedly hard, thanks in large part to the textbook, *The Feynman Lectures on Physics, Volume 1.* The book was less of a traditional textbook than a collection of brilliant essays based on a famous series of freshman physics lectures that Feynman delivered in the 1960s.

Unlike any other physics textbook that I have ever encountered,

The Feynman Lectures on Physics never bothers to explain how to solve any problems, which made trying to complete the daunting homework assignments challenging and time-consuming. What the essays did provide, however, was something much more valuable—deep insights into Feynman's original way of thinking about science. Generations have benefited from the *Feynman Lectures*. For me, the experience was an absolute revelation.

After a few weeks, I felt like my skull had been pried open and my brain rewired. I began to think like a physicist, and loved it. Like many other scientists of my generation, I was proud to adopt Feynman as my hero. I scuttled my original academic plans about biology and mathematics and decided to pursue physics with a vengeance.

I can remember a few times during my freshman year when I screwed up enough courage to say hello to Feynman before a seminar. Anything more would have been unimaginable at the time. But in my junior year, my roommate and I somehow summoned the nerve to knock on his office door to ask if he might consider teaching an unofficial course in which he would meet once a week with undergraduates like us to answer questions about anything we might ask. The whole thing would be informal, we told him. No homework, no tests, no grades, and no course credit. We knew he was an iconoclast with no patience for bureaucracy, and were hoping the lack of structure would appeal to him.

A decade or so earlier, Feynman had given a similar class, but solely for freshmen and only for one quarter per year. Now we were asking him to do the same thing for a full year and to make it available for all undergraduates, especially third- and fourth-year students like ourselves who were likely to ask more advanced questions. We suggested the new course be called "Physics X," the same as his earlier one, to make it clear to everyone that it was completely off the books.

Feynman thought a moment and, much to our surprise, replied

"Yes!" So every week for the next two years, my roommate and I joined dozens of other lucky students for a riveting and unforgettable afternoon with Dick Feynman.

Physics X always began with him entering the lecture hall and asking if anyone had any questions. Occasionally, someone wanted to ask about a topic on which Feynman was expert. Naturally, his answers to those questions were masterful. In other cases, though, it was clear that Feynman had never thought about the question before. I always found those moments especially fascinating because I had the chance to watch how he engaged and struggled with a topic for the first time.

I vividly recall asking him something I considered intriguing, even though I was afraid he might think it trivial. "What color is a shadow?" I wanted to know.

After walking back and forth in front of the lecture room for a minute, Feynman grabbed on to the question with gusto. He launched into a discussion of the subtle gradations and variations in a shadow, then the nature of light, then the perception of color, then shadows on the moon, then earthshine on the moon, then the formation of the moon, and so on, and so on, and so on. I was spellbound.

During my senior year, Dick agreed to be my mentor on a series of research projects. Now I was able to witness his method of attacking problems even more closely. I also experienced his sharp, critical tongue whenever his high expectations were not met. He called out my mistakes using words like "crazy," "nuts," "ridiculous," and "stupid."

The harsh words stung at first, and caused me to question whether I belonged in theoretical physics. But I couldn't help noticing that Dick did not seem to take the critical comments as seriously as I did. In the next breath, he would always be encouraging me to try a different approach and inviting me to return when I made progress.

One of the most important things Feynman ever taught me was that some of the most exciting scientific surprises can be discovered in everyday phenomena. All you need do is take the time to observe things carefully and ask yourself good questions. He also influenced my belief that there is no reason to succumb to external pressures that try to force you to specialize in a single area of science, as many scientists do. Feynman showed me by example that it is acceptable to explore a diversity of fields if that is where your curiosity leads.

One of our exchanges during my final term at Caltech was particularly memorable. I was explaining a mathematical scheme that I had developed to predict the behavior of a Super Ball, the rubbery, super-elastic ball that was especially popular at the time.

It was a challenging problem because a Super Ball changes direction with every bounce. I wanted to add another layer of complexity by trying to predict how the Super Ball would bounce along a sequence of surfaces set at different angles. For example, I calculated the trajectory as it bounced from the floor to the underside of a table to a slanted plane and then off the wall. The seemingly random movements were entirely predictable, according to the laws of physics.

I showed Feynman one of my calculations. It predicted that I could throw the Super Ball and that, after a complicated set of bounces, it would return right back to my hand. I handed him the paper and he took a glance at my equations.

"That's impossible!" he said.

Impossible? I was taken aback by the word. It was something new from him. Not the "crazy" or "stupid" that I had come to occasionally expect.

"Why do you think it's impossible?" I asked nervously.

Feynman pointed out his concern. According to my formula, if someone were to release the Super Ball from a height with a certain spin, the ball would bounce and career off nearly sideways at a low angle to the floor.

"And that's clearly impossible, Paul," he said.

I glanced down to my equations and saw that, indeed, my prediction did imply that the ball would bounce and take off at a low angle. But I wasn't so sure that was impossible, even if it seemed counterintuitive.

I was now experienced enough to push back. "Okay, then," I said. "I have never tried this experiment before, but let's give it a shot right here in your office."

I pulled a Super Ball out of my pocket and Feynman watched me drop it with the prescribed spin. Sure enough, the ball took off in precisely the direction that my equations predicted, scooting sideways at a low angle off the floor, exactly the way Feynman had thought was impossible.

In a flash, he deduced his mistake. He had not accounted for the extreme stickiness of the Super Ball surface, which affected how the spin influenced the ball's trajectory.

"How stupid!" Feynman said out loud, using the same exact tone of voice he sometimes used to criticize me.

After two years of working together, I finally knew for sure what I had long suspected: "Stupid" was just an expression Feynman applied to everyone, including himself, as a way to focus attention on an error so it was never made again.

I also learned that "impossible," when used by Feynman, did not necessarily mean "unachievable" or "ridiculous." Sometimes it meant, "Wow! Here is something amazing that contradicts what we would normally expect to be true. This is worth understanding!"

So eleven years later, when Feynman approached me after my lecture with a playful smile and jokingly pronounced my theory "Impossible!" I was pretty sure I knew what he meant. The subject of my talk, a radically new form of matter known as "quasicrystals," conflicted with principles he thought were true. It was therefore interesting and worth understanding.

Feynman walked up to the table where I had set up an experiment to demonstrate the idea. He pointed to it and demanded, "Show me again!"

I flipped the switch to start the demonstration and Feynman stood motionless. With his own eyes, he was witnessing a clear violation of one of the most well-known principles in science. It was something so basic that he had described it in the *Feynman Lectures*. In fact, the principles had been taught to every young scientist for nearly two hundred years . . . ever since a clumsy French priest made a fortuitous discovery.

PARIS, FRANCE, 1781: René-Just Haüy's face turned ashen, as the small sample of calcareous spar slipped out of his hands and fell to the floor with a crash. As he bent to collect the pieces, though, his sense of embarrassment melted away, replaced by curiosity. Haüy noticed that the surfaces where the sample had split apart were smooth and neatly angled, not rough and chaotic, as the outer surface of the original sample had been. He also noticed that the smaller pieces had facets that met at the same precise angles.

It was certainly not the first time someone had cracked open a rock. But this was one of those rare moments in history when an everyday occurrence leads to a scientific breakthrough because the person involved has both the instincts and the acumen to recognize the significance of what has just occurred.

Haüy had been born to humble beginnings in a French village. Early on, priests at a local monastery recognized his intellectual abilities and helped him achieve an advanced education. He eventually joined them in the Catholic priesthood and accepted a position teaching Latin at a Parisian college.

It was only after his theological career was under way that Haüy discovered his passion for the natural sciences. The turning point came when one of his colleagues introduced him to botany. Haüy was fascinated by the symmetry and the specificity of plants. Despite their tremendous variety, plants could be precisely classified on the basis of their color, shape, and texture. The thirty-eight-year-old priest soon became an expert in the subject, frequently visiting the Jardin du Roi in Paris to test his identification skills.

Then, during one of his many visits to the Jardin, Haüy was exposed to another field of science that was to become his true calling. The great naturalist Louis-Jean-Marie Daubenton had been invited to give a public lecture about minerals. During the presentation, Haüy learned that minerals, like plants, come in many different colors, shapes, and textures. But at that point in history, the study of minerals was a much more primitive discipline than botany. There was no scientific classification of the various types of minerals nor any understanding about how they might be related to one another.

Scientists knew that minerals, like quartz, salt, diamond, and gold, are solely composed of one pure substance. If you were to smash them to bits, each bit would consist of exactly the same material. They also knew that many minerals form faceted crystals.

But unlike plants, two minerals of the same type can have very different colors, shapes, and textures. Everything depends on the conditions under which they are formed and what happens to the mineral afterward. In other words, minerals seemed to defy the neat and tidy classification that Haüy had come to appreciate about botany.

The lecture inspired him to ask an acquaintance, the wealthy financier Jacques de France de Croisset, if he could examine his private mineral collection. The visit was a joy for Haüy, up until the fateful moment when he dropped the sample of calcareous spar.

The financier graciously accepted Haüy's apologies for the damage he had caused. But he also noticed his guest's absolute fixation on the shattered remains and generously offered to let him take some of the pieces home for further study.

Back in his room, Haüy took a small fragment of an irregular shape and carefully cleaved its surfaces, chipping away, bit by bit, until the exterior consisted entirely of smooth, flat facets. He noticed that the facets formed a small rhombohedron, the relatively simple shape of a cube pushed on an angle.

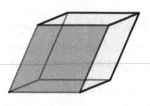
Rhombohedron

Haüy then took another calcareous spar fragment with a rough outer shape and repeated the same operation. Once again a rhombohedron emerged. This time the size of the rhombohedron was somewhat larger, but it had the exact same angles as the one he had tested before. Haüy repeated the experiment many times, utilizing all of the different fragments he had been given. Later he did the same for many other samples of calcareous spar found in different regions of the world. Each time he found the same result: a rhombohedron with the same angles between facets.

The simplest explanation Haüy could think of was that the calcareous spar was composed of a basic building block that was, for some unknown reason, shaped like a rhombohedron.

Haüy then expanded his experiments to include other types of minerals. In each case, he found that the mineral could be cleaved and reduced to a building block with a certain precise geometrical shape. Sometimes it was a rhombohedron, just like the calcareous spar. Sometimes it was a rhombohedron with different angles between facets. Sometimes it was a different shape altogether. He shared some of his findings with French naturalists and won broad acceptance from the scientific community, which enabled him to continue his methodical study of minerals for the next two decades, including throughout the French Revolution.

Haüy finally published his masterpiece, the *Traité de Minéralogie*, in 1801. It was a superbly illustrated atlas compiling his results and presenting the "laws of crystal forms" that he had discovered while gathering his data.

The publication was an instant classic. It earned him an academic scientific position, the admiration of his peers, and a place in history as the "Father of Modern Crystallography." Haüy's scientific contributions were considered so important that Gustav Eiffel chose to include him on the list of seventy-two French scientists, engineers, and mathematicians whose names are engraved on the first floor of the Eiffel Tower.

A profound implication of Haüy's work was that minerals are composed of some kind of primitive building block, which he called *la molécule intégrante*, that repeats over and over throughout the material. Minerals of the same type are constructed from the same building block no matter where in the world they may originate.

Several years later, Haüy's discovery helped inspire an even bolder idea. British scientist John Dalton proposed that all matter, not just minerals, is made of indivisible and indestructible units called atoms. According to this idea, Haüy's primitive building blocks corresponded to a cluster of one or more atoms whose type and spatial arrangement determined the type of mineral.

The ancient Greek philosophers Leucippus and Democritus are often credited with introducing the concept of atoms in the fifth century BCE. But their ideas were strictly philosophical. It was Dalton who transformed the atomic hypothesis into a testable scientific theory.

From his experience studying gases, Dalton concluded that atoms are spherical in shape. He also proposed that different types of atoms have different sizes. They were far too small to be seen by cleaving minerals or using any of the other technologies available in the nineteenth century. So it would take more than a century of fierce debate and the development of new technologies and new types of experiments before the atomic theory was fully accepted.

Despite their accomplishments, neither Haüy nor Dalton could explain one of Haüy's most important discoveries. No matter which mineral he studied, the primitive building block, *la molécule intégrante*, was either a tetrahedron, a triangular prism, or a parallelepiped, which is a broader category that includes the rhombohedron that Haüy originally observed. Why should that be so?

The search for an explanation continued for many decades, ultimately leading to the creation of a new and pivotal field of science known as "crystallography." Based on rigorous mathematical principles, crystallography would eventually make an enormous impact on other scientific disciplines, including physics, chemistry, biology, and engineering.

The laws of crystallography would turn out to be powerful enough to explain all of the possible forms of matter known at the time and to predict many of their physical properties, such as hardness, response to heating and cooling, conduction of electricity, and elasticity. Crystallography's success in explaining so many different properties of matter relevant to so many different disciplines has long been considered one of the great triumphs of nineteenth-century science.

Yet, by the early 1980s, it was precisely these celebrated laws of crystallography that my student Dov Levine and I were challenging.

Tetrahedron

Triangular prism

Parallelepiped

We had figured out how to construct novel building blocks that could be packed into arrangements that were supposedly impossible. The fact that we had discovered something new in what was thought to be a well-understood, fundamental principle of science was what had grabbed Feynman's attention during my lecture.

To fully appreciate his surprise warrants a brief introduction to the three simple principles that are at the foundation of crystallography:

The first principle is that all pure substances, such as minerals, form crystals, as long as there is enough time for the atoms and molecules to move into an orderly arrangement.

The second states that all crystals are periodic arrangements of atoms, meaning their structure is entirely composed of one of Haüy's primitive building blocks, a single cluster of atoms that repeats over and over in any direction with equal spacing.

The third principle is that every periodic atomic arrangement can be categorized according to its symmetries, and there is a finite number of possible symmetries.

This third principle is the least obvious of the three, but can be easily illustrated with everyday floor tiles. Imagine that you want to cover a floor with regularly spaced tiles that have identical shapes, as seen in the examples opposite. Mathematicians call the resultant pattern a periodic tiling. The tiles are two-dimensional analogs of Haüy's three-dimensional primitive building blocks because the entire pattern is composed of repeated elements of the same unit. Periodic tilings are frequently used in kitchens, patios, bathrooms, and entryways. And those patterns often include one of five basic shapes: rectangles, parallelograms, triangles, squares, or hexagons.

But what other simple shapes are possible? Stop and think about this for a moment. What other basic shape could you use to tile your floor? How about a regular pentagon, a five-sided shape whose edges have equal length and whose angles have equal measure?

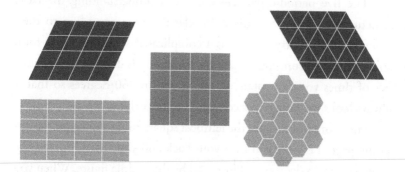

The answer may surprise you. According to the third principle of crystallography, the answer is no. Absolutely not. A pentagon won't work. In fact, nothing else works. Every two-dimensional periodic pattern corresponds to one of the five patterns shown above.

You might find a floor's tiling pattern that seems to be an exception to the rule. But that is a bit of trickery. If you take a closer look, the tiling will always turn out to be one of the same five patterns in disguise. For example, you could make more complicated-looking patterns by replacing each of the straight edges with identical curvy ones. You could also cut or divide each tile—for example, a square along the diagonal— and then fit them back into a pattern using other geometric shapes. Or you could choose a picture or design and insert it in the center of each tile. But, from a crystallographer's point of view, none of that changes the fact that the framework is equivalent to one of the five possibilities above. No other fundamental patterns exist.

If you asked your contractor to cover your shower floor with regular pentagons, you would actually be asking for a lot of water damage. No matter how the tiler tried to jam the pentagons together, there would always be gaps (see the next page). Lots of them! The same would be true if you tried a regular seven-sided heptagon, eight-sided octagon, or nine-sided nonagon. The list of forbidden shapes goes on and on forever.

The five periodic patterns are key to understanding the basic structure of matter. Scientists also classify them according to their "rotational symmetry," which is a complicated-sounding name for a straightforward concept. Rotational symmetry is defined as the number of times you can rotate an object within 360 degrees so that it always looks unchanged, as compared to the original.

For example, consider the leftmost square tiling shown on the opposite page. Let's say you turn your back and your friend rotates the square tiling 45 degrees, as illustrated by the middle figure. When you return to face the tiling, you are able to see it does not look the same as it did originally, being obviously oriented in a different direction. So this rotation of 45 degrees is not considered a "symmetry" of the square.

But, starting over again, if your friend rotates the tiling 90 degrees (the right-hand figure on the facing page) then you will *not* be able to detect that anything has changed. The tiling looks exactly the same as

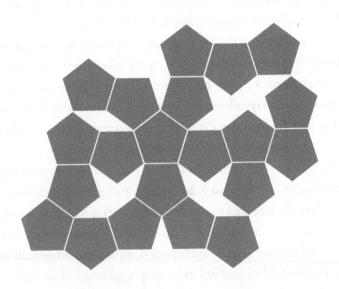

it did originally. This rotation by 90 degrees is considered a rotational "symmetry." In fact, 90 degrees is the minimal angle of rotation that creates a symmetry of the square pattern. Any rotation of the square less than 90 degrees alters the apparent orientation.

It then becomes clear that two rotations by 90 degrees, for a total of 180 degrees, is also a symmetry. So are three rotations (270 degrees) and four (360 degrees). Since it takes a total of four rotations to complete 360 degrees, the square tiling is said to have four-fold symmetry.

Let's give your friend a tiling composed of equal rows of rectangles whose long sides are oriented horizontally. Rotating the tiling by 90 degrees makes it look different because the long sides are now oriented vertically. But rotating it 180 degrees makes it looks the same as it did originally. So in the case of a rectangle, 180 degrees is the smallest rotation that is a symmetry. Twice that rotation is 360 degrees. So a tiling of rectangles has a two-fold symmetry.

Similarly, for a tiling of parallelograms, the only rotation that leaves the tiling looking unchanged is 180 degrees. Therefore, a parallelogram tiling also has two-fold rotational symmetry.

Using the same method, an equilateral triangle can be seen to have three-fold symmetry. A hexagon has six-fold.

Finally, there is one other possible rotational symmetry that can

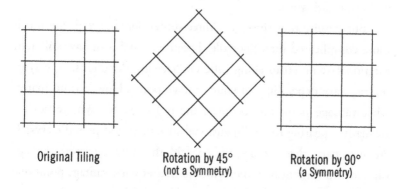

Original Tiling   Rotation by 45°   Rotation by 90°
                         (not a Symmetry)   (a Symmetry)

be made from any of the five patterns. If we make the edges of any one of the shapes irregularly jagged, for example, the only rotation that leaves the pattern looking unchanged would be a complete turn of 360 degrees, or one-fold symmetry.

And that completes the list of possibilities. One-, two-, three-, four-, and six-fold are the only rotational symmetries allowed for two-dimensional periodic patterns, a fact that has been known for millennia. Ancient Egyptian artisans, for example, used rotational symmetries to create beautiful mosaics. But it was not until the nineteenth century that those trial-and-error methods were fully explained by rigorous mathematics.

But let's get back to your shower floor. The fact that your contractor cannot make a periodic pattern using only regular five-sided pentagon tilings without leaving substantial gaps and creating water damage is a vivid demonstration that five-fold symmetry is impossible according to the laws of crystallography. But it is not the only forbidden symmetry. The same is true for seven-, eight-, and any higher-fold symmetry.

Remember that Haüy discovered that crystals are periodic, just like the tiles on your floor, with regularly repeating patterns. So by extension, the same types of restrictions that apply to tilings also apply to three-dimensional crystals. Only certain patterns can fit together without creating gaps.

But despite that similarity, three-dimensional crystals are much more complicated than floor tiles because crystals can have different rotational symmetries along different viewing directions. The symmetries vary depending on your vantage point. However, no matter what vantage point one chooses, the only possible symmetries for regularly repeating three-dimensional structures and periodic crystals are one-, two-, three-, four-, and six-fold, the same restriction that applies to two-dimensional tiles. And no matter what vantage point one chooses, *five-fold rotational symmetry is always forbidden*, along with seven-, eight-, and any higher-fold symmetry.

How many distinct combinations of symmetries as viewed along all possible vantage points are possible for periodic crystals? Finding the answer was a great mathematical challenge.

The problem was finally solved in 1848 by French physicist Auguste Bravais, who showed that there are exactly fourteen distinct possibilities. Today, these are known as the Bravais lattices.

But the challenge to understanding crystal symmetries did not end there. A more complete mathematical classification was later developed combining rotational symmetries with even more complicated symmetries, known as "reflections," "inversions," and "glides." When all of these additional possibilities are added to the mix, the number of possible symmetries altogether grows from 14 to a total of 230. But among all of those possibilities, five-fold symmetry remains forbidden along any direction.

These discoveries brought the beauty of mathematics together with the beauty of the natural world in a most remarkable way. The identification of all of the 230 possible three-dimensional crystal patterns was accomplished using pure mathematics. And each of those patterns could also be found in nature by cleaving minerals.

The remarkable correspondence between the abstract mathematical crystal patterns and the real crystals found in nature was indirect but compelling evidence that matter is composed of atoms. But exactly how were those atoms arranged? Cleaving could reveal the shape of the building blocks, but it was far too crude of a tool to determine how atoms are arranged within.

A precise tool capable of obtaining this information was invented by German physicist Max von Laue at the University of Munich in 1912. He discovered that he could precisely determine the hidden symmetry of a chunk of matter simply by shining a beam of X-rays through a small sample of the material.

X-rays are a type of light wave whose wavelength is so small that they can easily pass through the channels of empty space between the regularly spaced rows of atoms in crystals. When the X-rays passing through the crystal are then projected onto a piece of photographic paper, von Laue showed, they interfere with one another to produce a pinpoint pattern of sharp spots known as an "X-ray diffraction pattern."

If the X-rays are pointed along a line of rotational symmetry through the crystal, the pinpoint diffraction pattern has precisely the same symmetry. By shining the X-rays through a crystal along various directions, the full set of symmetries of its atomic structure can be revealed. And from that information, the crystal Bravais lattice and the shape of its building block can also be identified.

Shortly after von Laue's discovery, the father and son team of British physicists, William Henry and William Lawrence Bragg, took the next big step. By carefully controlling the X-ray wavelength and direction, they showed that the pinpoint diffraction pattern could not only be used to reconstruct the symmetry, but also the detailed atomic arrangement throughout the crystal. The pinpoints became known as "Bragg peaks."

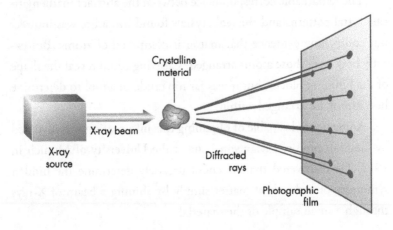

The two breakthroughs immediately became indispensable in the exploration of matter. Over subsequent decades, tens of thousands of diffraction patterns were obtained from various natural and synthetic materials from all over the world. In later years, scientists obtained even more accurate information by replacing X-rays with electrons, neutrons, or the high-energy radiation produced when a beam of charged particles moving at relativistic speeds is bent at an angle by the magnets in a powerful particle accelerator called a synchrotron. But no matter the method, the original rules of symmetry derived from Haüy's and Bravais's work were always obeyed.

Based on the combination of mathematical reasoning and accumulated experimental experience, the rules became firmly fixed in the minds of scientists. It seemed certain, or at least as certain as any scientific principle could ever be, that matter could only have one of the long-prescribed symmetries. Nothing else. Five-fold symmetry remained verboten for more than two hundred years.

PASADENA, 1985: But now, here I was, standing in front of Richard Feynman explaining that these long-standing rules were wrong.

Crystals were not the only possible forms of matter with orderly arrangements of atoms and pinpoint diffraction patterns. There was now a vast new world of possibilities with its own set of rules, which we named quasicrystals.

We chose the name to make clear how the new materials differ from ordinary crystals. Both materials consist of groups of atoms that repeat throughout the entire structure.

The groups of atoms in crystals repeat at regular intervals, just like the five known patterns. In quasicrystals, however, different groups repeat at distinct intervals. Our inspiration was a two-dimensional pattern known as a Penrose tiling, which is an unusual pattern that

contains two different types of tiles that repeat at two incommensurate intervals. Mathematicians call such a pattern quasiperiodic.
Hence, we dubbed our theoretical discovery "quasiperiodic crystals"
or "quasicrystals," for short.

My little demonstration for Feynman was designed to prove my
case using a laser and a slide with a photograph of a quasiperiodic
pattern. I flipped on the laser, as Feynman had directed, and aimed
the beam so that it passed through the slide onto the distant wall.
The laser light produced the same effect as X-rays passing through the
channels between atoms: It created a diffraction pattern, like the one
pictured in the photo below.

I turned off the overhead lights so that Feynman could get a good
look at the signature snowflake pattern of pinpoints on the wall. It
was unlike any other diffraction pattern that Feynman had ever seen.

I pointed out to him, as I had done during the lecture, that the
brightest spots formed rings of ten that were concentric. That was

unheard of. One could also see groups of pinpoints that formed pentagons, revealing a symmetry that was thought to be absolutely forbidden in the natural world. A closer look revealed yet more spots between the pinpoints. And spots between those spots. And yet more spots still.

Feynman asked to look more closely at the slide. I switched the lights back on and removed it from the holder and gave it to him. The image on the slide was so reduced that it was hard to appreciate the detail, so I also handed him an enlargement of the tiling pattern, which he put down on the table in front of the laser.

The next few moments passed in silence. I began to feel like a student again, waiting for Feynman to react to the latest cockamamie idea I had come up with. He stared at the enlargement on the table, reinserted the slide in the holder, and switched on the laser himself. His eyes went back and forth between the printed enlargement on the table, up to the laser pattern on the wall, then back down again to the enlargement.

"Impossible!" Feynman finally said. I nodded in agreement and smiled, because I knew that to be one of his greatest compliments.

He looked back up at the wall, shaking his head. "Absolutely impossible! That is one of the most amazing things I have ever seen."

And then, without saying another word, Dick Feynman looked at me with delight and gave me a huge, devilish smile.

TWO

THE PENROSE PUZZLE

PHILADELPHIA, PENNSYLVANIA, OCTOBER 1981: Four years before my encounter with Feynman, no one had ever heard of quasicrystals. Including me.

I had just joined the Department of Physics at the University of Pennsylvania and was invited to give the Physics Colloquium, a weekly lecture attended by the entire department. Penn had recruited me to the faculty based on my work at Harvard University in elementary particle physics, which was related to understanding the fundamental constituents of matter and the forces through which they interact. There was also great interest in my most recent research. My first graduate student, Andy Albrecht, and I were working feverishly on developing novel ideas about the creation of the universe, ideas that would eventually help set the foundation for what is now known as the inflationary theory of the universe.

But I decided not to talk about any of that. Instead, I chose to talk about a project that almost no one knew I had been working on and whose significance was not yet clear. I did not know the lecture would resonate with a young graduate student who was sitting in the audience, or that it would soon lead to a fruitful partnership and the discovery of a new form of matter.

Most of my presentation described a project that I had been exploring for the last year and a half with David Nelson, a theoretical physicist at Harvard, and Marco Ronchetti, a postdoctoral fellow working at the IBM Thomas J. Watson Research Center in Yorktown Heights, New York.

Our project was to study how atoms in a liquid rearrange themselves if the liquid is rapidly cooled and solidified. It was well known among scientists that when a liquid is frozen very slowly, its atoms tend to reconfigure from the random arrangement of a liquid into the orderly, periodic arrangement of a crystal (as when water freezes into ice).

For the simplest case, in which all the atoms are identical and interacting under simple interatomic forces, the atomic arrangement would be one in which the atoms stacked together like oranges on display at a grocery store. The structure—technically called face-centered cubic—has the same symmetry as a cube, consistent with all the known rules of crystallography.

The three of us wanted to study what would happen if the liquid was cooled so rapidly that it solidified before the atoms had a chance to rearrange themselves into a perfect crystal. The common scientific assumption at the time was that the atomic arrangement would be like a snapshot of the liquid state. In other words, it would be entirely random, with no discernible order.

David Nelson and one of his students, John Toner, had conjectured that something more subtle might happen. Rapid solidification could result in a mixture of randomness and order. The atoms would be randomly placed in space, but the bonds between those atoms could align, on average, along the edges of a cube, they theorized. The atomic order would then be somewhere between order and disorder. They called the special phase "cubatic."

To understand the significance of that idea, one must first understand a few basics. The physical properties of matter and how it can be used depend critically on the configuration of its atoms and molecules. For example, consider the crystals graphite and diamond. Based on their physical properties, it is hard to imagine that the two have anything in common. Graphite is soft, slippery, and opaque with a dark metallic appearance. Diamond is ultra-hard, transparent, and shiny. However, both are composed of exactly the same types of atoms and are 100 percent carbon. The only difference between the two substances is how the carbon atoms are arranged, as illustrated below.

In diamond, each carbon atom is bonded to four other carbon atoms in an interconnected three-dimensional network. In graphite, each carbon atom is bonded to only three other carbon atoms in a two-dimensional sheet. The sheets of carbon stack together, one atop the other, like sheets of paper.

The diamond's network is sturdy and difficult to break apart. Sheets of carbon, on the other hand, easily slip past one another like pieces of paper. That is the basic reason why diamond is so much stronger than graphite. And that difference has a direct effect on their practical applications. Diamond, one of the hardest materials known, is used for drill bits. Graphite, on the other hand, is so soft it is used for pencils. Sheets of carbon peel off as the pencil moves across the page.

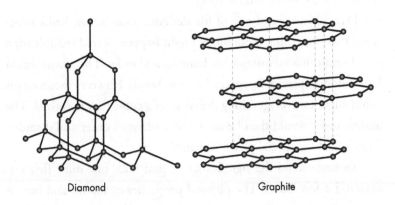

Diamond   Graphite

This example illustrates how knowing the symmetry of an atomic arrangement in a material makes it possible to understand and predict its properties and to figure the most effective uses for it. The same applies to rapidly cooled solids, which scientists call glassy or amorphous. They are a valuable alternative to slowly cooled crystals because they are observed to have different electronic, thermal, elastic, and vibrational properties. Slowly cooled crystal silicon, for example, is used extensively throughout the electronics industry. But amorphous silicon is less rigid than the slowly cooled material, which makes it advantageous in certain types of solar cells.

The issue that Nelson, Ronchetti, and I wanted to investigate was whether some rapidly cooled solids have a subtle kind of order that had not been detected before and that might suggest additional advantages and applications.

I had already been working for several years on developing ways to simulate the rapid cooling of liquids. I had been invited to spend my summers, first as an undergraduate and then again as postdoctoral fellow, to work on theoretical computer models at Yale University and the IBM Thomas J. Watson Research Center. My main scientific interests were pointing elsewhere at the time. But I took advantage of the summer research opportunities because I was intrigued by the fact that the atomic arrangement of something as rudimentary as amorphous matter was not yet known. In that, I was intentionally following one of the most important lessons learned from my mentor, Richard Feynman: It is wise to follow your heart and seek out good problems wherever they may lead, even if it is not in the direction you thought you were supposed to be headed.

I developed the first computer-generated continuous random network (CRN) model of glass and amorphous silicon in 1973, the summer before my senior year at Caltech. The model was widely used to predict structural and electronic properties of these materials. In later years, while working with Ronchetti, I developed more sophisticated programs to simulate the rapid cooling and solidification process.

In 1980, a chance conversation at Harvard with David Nelson gave new purpose to all my efforts on amorphous material. My computational models could be adapted to test Nelson and Toner's speculation about cubatic matter.

After explaining all of the background history to the audience at Penn, I moved toward the climax of the talk: If the conjecture about a cubatic phase was right, the atomic bonds in my new computer simulations should not be oriented randomly. On average, the bonds should tend toward a "cubic orientation," preferentially aligned along the edges of a cube.

We developed a sophisticated mathematical test for the experiment to check whether the average orientations of the bonds displayed the expected cubic symmetry, assigning a numerical score according to how strong the cubic alignment was.

And the result was . . . utter failure. We found no sign whatsoever of a preferential alignment of bonds along the edges of a cube, which Nelson and John Toner had predicted.

But by accident, we discovered something even more interesting. While devising a quantitative mathematical test to check for the orientation of the atomic bonds with the symmetry of a cube, we found that it would be easy to adapt the test to scan for every other possible rotational symmetry. So, as an afterthought, we used the test to give each symmetry a score, based on the degree of alignment among the atomic bonds along different directions.

To our great surprise, a forbidden symmetry scored much higher than the rest—the impossible symmetry of an icosahedron, shown in the left figure on the facing page.

I knew that some in the audience would already be familiar with an icosahedron because the three-dimensional shape was being used as a die in the popular game *Dungeons & Dragons*, seen below. Others would recognize it from biology, where the shape appears in certain human viruses. The more geometrically inclined would recognize it as one of the five Platonic solids, three-dimensional shapes in which every face is identical to every other, every edge is identical to every other, and every corner is identical to every other.

The significant feature of a three-dimensional icosahedron is that, staring directly down at any of the corners, one observes a pentagonal shape with five-fold symmetry. The same five-fold symmetry that is forbidden for two-dimensional tilings or three-dimensional crystals.

Of course, there is nothing wrong with having a single tile in the shape of a regular pentagon. You can make a single tile any shape you wish. But it is impossible to cover a floor with regular pentagons without leaving gaps. The same applies to an icosahedron. It is possible to make a single three-dimensional die in the shape of an icosahedron. But you cannot fill space with icosahedrons without leaving gaps and holes, as illustrated by the image on the next page.

With so many corners, each with forbidden five-fold symmetry,

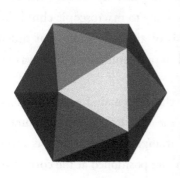

the icosahedron was well known among scientists studying the struc-ture of matter to be the most forbidden symmetry for an atomic ar-rangement. This fact was considered so basic that it often appeared in chapter 1 in textbooks. Yet somehow, icosahedral symmetry had received the highest score for the alignment of atomic bonds in our computer experiment.

Strictly speaking, our results had not directly violated the laws of crystallography. Those rules only apply to macroscopic chunks of matter containing tens of thousands of atoms or more. For much smaller groups of atoms, as studied in our simulation, there was no absolute restriction.

In the extreme case of a small cluster containing only thirteen identical gold atoms, for example, the interatomic forces naturally move the atoms into an icosahedral arrangement. One atom is at the center and twelve surrounding atoms are positioned at the corners of an icosahedron. That occurs because interatomic forces are springlike and tend to draw atoms together into tightly packed symmetrical arrangements. The icosahedron occurs for thirteen atoms because it is the most symmetrical tightly-packed configuration that can be achieved. However, as more and more atoms are added, icosahedral symmetry becomes less favored. As shown on the opposite page with the *Dungeons & Dragons* die, icosahedrons cannot fit neatly together face-to-face, edge-to-edge, or in any other way that does not leave large gaps between them.

What was surprising about our calculation was that the icosahedral symmetry of the bond orientations extended nearly all the way across a simulation containing thousands of atoms. If you had asked most experts at the time, they would have guessed that icosahedral symmetry could not possibly extend over more than fifty atoms or so. But our simulations showed that there remained a high degree of icosahedral symmetry among the bond orientations even when averaged over many atoms. The laws of crystallography, however, demanded that icosahedral symmetry not extend indefinitely. And, sure enough, when we continued to average over yet more and more atoms, the symmetry score began to drop and eventually reached a level that was no longer statistically significant. Even so, the discovery of a high degree of bond alignment along the edges of an icosahedron for thousands of atoms was something truly remarkable.

I reminded the audience that the icosahedral ordering had arisen spontaneously from simulations containing only *one* type of atom. Most materials contain a combination of different elements with different sizes and different bonding forces. With an increasing number of different elements, I hypothesized, it might become easier to violate the known rules of crystallography so that icosahedral symmetry could extend over an increasing number of atoms.

Perhaps there could even be circumstances in which the symmetry extended without limit, I suggested. That would be nothing short of revolutionary, a direct violation of the laws Haüy and Bravais established more than a century ago. It was the first time I had expressed such an impossible idea in public, and I ended my lecture with that provocative thought.

There was an enthusiastic round of applause. Several faculty members asked me questions about this detail or that. And I received a lot of nice compliments afterward. But no one commented on my wild speculation about violating the laws of crystallography. Perhaps they all assumed it was merely a rhetorical flourish.

There was one person in the audience, though, who took me seriously. And he was about to gamble his entire future on the idea. The day after my talk, a twenty-four-year-old physics graduate student named Dov Levine showed up at my office and asked if I would be his new PhD advisor. Dov was specifically interested in working with me on the crazy idea he had heard at the end of my talk.

My initial reaction was not very encouraging. "That idea is nuts," I told him. I would never recommend that kind of problem to a graduate student, I warned. I was not even sure I would recommend it to an untenured professor like myself. I had only a vague notion about where to start, and the chance of success was laughably small. I went on and on making an endless series of discouraging remarks, but nothing I said seemed to faze him. Dov was emphatic that he wanted to give it a go, no matter the odds.

When I asked Dov to tell me more about himself, he began by explaining that he was born and raised in New York City. That much was already obvious to me, based on his rapid cadence, brash attitude, and wry sense of humor. Dov could never go three sentences without making a joke or an irreverent remark, always accompanied by a mischievous smile.

I tried not to reveal what I was thinking as I listened to Dov argue why we should pursue my wild idea. But I secretly approved that he was apparently a stubborn person who was not easily dissuaded. That is the kind of attitude one needs to take on a super-high-risk problem, I thought. A good sense of humor would come in handy, as well, since we were likely to encounter more than a few difficulties.

There was something else working in Dov's favor—a dream of mine that stretched back to when I was thirteen and read the novel *Cat's Cradle,* by Kurt Vonnegut. The book is about the potential misuse of science, which was admittedly a strange novel to inspire a budding scientist.

In the book, Vonnegut imagined a new form of frozen water called "ice-nine." When a seed crystal of ice-nine makes contact with ordinary water, it causes all of the H_2O molecules to rearrange themselves into a solid block. A single seed crystal, if thrown in the ocean, could trigger a chain reaction and solidify all of the water on the planet.

Ice-nine was, of course, a fictional creation. But the novel brought to my attention a scientific fact that I had never considered before, which was that the properties of matter can be radically changed by simply rearranging its atoms.

Maybe, just maybe, I thought, there were other forms of matter whose arrangements of atoms had not yet been observed by scientists. And maybe, I imagined, they did not even occur on this planet.

Dov had no way of knowing it, but he was about to help provide me with the opportunity to pursue my longtime scientific fantasy. I agreed to take him on as a student on a trial basis. If we made no progress after six months, we both understood that he might have to find a different topic and a different advisor.

We began by trying to determine the largest number of atoms we could place in a tightly packed arrangement with the symmetry of an icosahedron. In order to visualize what we were doing,

Dov, as seen on the left, and I needed to construct some kind of tangible model. But there, we immediately ran into a roadblock. Chemists constructed such models using commercially available kits containing plastic spheres and rods. That was fine, as long as one was studying ordinary crystalline arrangements.

Dov and I were trying to do something different. We needed pieces that could produce bond angles and bond distances appropriate for the symmetry of an icosahedron. Since the symmetry was impossible for crystals, chemistry kits did not include such pieces. Everyone, including the model makers, knew that five-fold symmetry was forbidden. So we had to improvise, and finally resorted to experimenting with Styrofoam balls and pipe cleaners. Before too long, my office started looking like an arts-and-crafts project gone berserk.

We began by assembling a cluster of thirteen Styrofoam balls into the shape of an icosahedron, like the one I had described in my Penn lecture, with one ball in the middle and the other twelve lying at the corners of an icosahedron, as shown on the opposite page.

Then we tried to surround this first icosahedron with twelve more identical icosahedrons, constructing a larger, more complex structure—an "icosahedron of icosahedrons." But that created an immediate problem. The icosahedrons did not fit together very well.

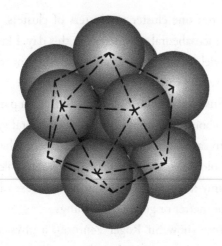

There were large gaps in between them. So we tried to preserve the structure by adding more Styrofoam balls and more pipe cleaners to fill all of the open spaces between the individual icosahedrons. That method worked well enough for us to build a large cluster with the symmetry of an icosahedron containing more than two hundred atoms.

We then tried to repeat our success, this time using thirteen copies of this large cluster to build an even larger cluster. But the gaps that we wound up creating were much larger now, and the model kept falling apart.

Our simple arts-and-crafts project appeared to illustrate a fundamental limitation in creating atomic structures with icosahedral symmetry. Because individual icosahedrons do not fit together neatly, there would always be ever larger gaps to fill as more atoms were added to the structure. From this experience, we hypothesized that it would be impossible to extend icosahedral symmetry beyond a few hundred, or perhaps a few thousand, atoms.

Dov and I wrongly assumed that our strategy of building

hierarchically, from one cluster to clusters of clusters, was the only way to maintain icosahedral symmetry. To this day, I keep one of the pipe cleaner models in my office as a reminder of how close we came to reaching the wrong conclusion.

The two of us were thinking about publishing a paper describing our conclusion about the impossibility of icosahedral symmetry. But Dov managed to save us from that embarrassment when he brought me a four-year-old *Scientific American* article about Penrose tilings. *Penrose?* I certainly knew the name. But not because of anything related to forms of matter or geometrical tilings.

Roger Penrose (now Sir Roger Penrose), a physicist at Oxford University, was already recognized worldwide for his many contributions to general relativity and its applications to understanding the evolution of the universe. In the 1960s, Penrose proved a set of influential "singularity" theorems showing that, under a wide range of conditions, a universe that is expanding today must have emerged from a big bang. More than four decades later, some cosmologists, including me, are considering ways of avoiding those initial conditions in order to avoid the big bang and replace it with a big bounce.

As luck would have it, the only reason Dov knew about Penrose tilings was because he had originally come to Penn to work on general relativity. In December 1980, a year before he attended my lecture, he had heard Penrose talking about his tiling patterns at an international conference.

BALTIMORE, MARYLAND, 1980: Dov was attending the "Tenth Texas Symposium on Relativistic Astrophysics." It was an odd name for the convention, because Baltimore is more than a thousand miles from Texas. The name follows an informal tradition. Texas was home to the first Symposium on Relativistic Astrophysics, so every meeting thereafter keeps the original name, even if it is held in Geneva, Switzerland.

Dov was strolling down the halls in between scientific talks when he happened to overhear someone chatting with a group of students about some new work by Roger Penrose. Thinking this would be about relativity, he moved closer to eavesdrop on the conversation.

To Dov's surprise, the discussion was not about general relativity. Instead, it was about a novel tiling that Penrose had constructed a few years earlier for his own amusement. He basically discovered it by doodling. Penrose made sketches of tiles, and groups of tiles, in his notebook until he came up with a tiling that could solve a famous mathematical puzzle. In addition to being a creative genius with unbounded curiosity, Penrose was also an extraordinarily talented artist who could draw precise figures freehand. Throughout his career, Penrose has often used his intricate hand-drawn illustrations to clarify highly technical points in his seminars.

Inventing a new type of tiling may seem like an odd form of amusement. For Penrose, it was an exercise in "recreational mathematics," a pastime that entails exploring certain well-known mathematical puzzles and challenges. Aficionados range from total amateurs to famous mathematicians, and from young to old.

The leading exponent of recreational mathematics at the time was Martin Gardner, who wrote a monthly column in *Scientific American* for twenty-five years entitled "Mathematical Games."

The article Dov brought me was Martin Gardner's *Scientific American* article about Penrose tilings published in 1977, about three years after Penrose had invented the tilings. The article explained how Penrose had found a neat solution to a challenge that recreational mathematicians had been discussing for many years: Is it possible to find a set of tiles that can cover a floor without leaving gaps and do so *only nonperiodically?*

Triangles can cover a floor nonperiodically if, for example, they are arranged in a spiral, as shown by the pattern in the left illustration below. However, triangles can also make patterns that are periodic, as shown in the right illustration below. So triangles would not be a valid solution to the challenge.

Mathematicians once thought it was impossible to find any shape or combination of shapes that could satisfy the challenge. But in 1964, mathematician Robert Berger constructed a valid example composed of 20,426 different tile shapes. Over the years, others managed to find examples using many fewer tile shapes.

In 1974, Penrose made a major breakthrough when he found a solution to the challenge using only *two* tile shapes, which he called "kites" and "darts" (see opposite page). Each of the tiles is marked with circular arcs or "ribbons." Penrose imposed a rule that two tiles can only be joined together edge-to-edge if the ribbons on both sides of the joint edge match. Following this "matching rule" prevents the tiles from being put together in any regularly repeating pattern. The tiling on the opposite page shows the complex ribbon pattern that emerges when many kites and darts are put together following Penrose's matching rules.

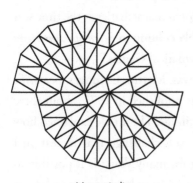

Nonperiodic

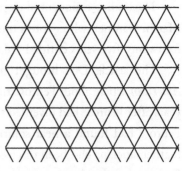

Periodic

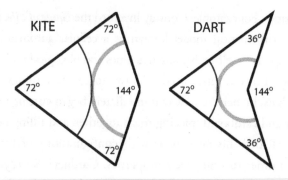

Look closely to pick out the kites and darts.

PHILADELPHIA, OCTOBER 1981: The Gardner article described many surprising features of the original tilings Penrose had discovered, as well as additional properties subsequently discovered by his friend, Cambridge University mathematician John Conway.

Conway has made countless contributions to number theory, group theory, knot theory, game theory, and other fundamental fields of mathematics. For example, Conway invented the *Game of Life*, a famous abstract mathematical model known as a cellular automaton, which mimics aspects of self-replicating machines and biological evolution.

When Penrose introduced Conway to the new tilings, he went wild with excitement. Conway immediately began cutting out pieces of paper and cardboard, piecing them together, and filling tables and surfaces all over his apartment with configurations of his cut-out shapes in order to study their properties. Gardner's *Scientific American* article included many of Conway's valuable insights that helped Dov and me flesh out certain properties of Penrose tilings that were not obvious at first sight.

In reading other articles, we learned that the precise shapes of the tiles were not important so long as the tiles fit together in ways that were equivalent to kites and darts. A version that was easier for Dov and me to analyze was constructed from a pair of fat and skinny rhombuses, the four-sided shapes used to construct the tiling pattern seen at the top of the opposite page.

It is possible to arrange just the fat rhombuses into a periodic pattern, or just the skinny shapes into a periodic pattern, or to arrange many different combinations of the two shapes together into various other periodic patterns.

But the rhombuses are not the whole story. To prevent all the periodic possibilities and *force* a nonperiodic arrangement, it is necessary to introduce some sort of matching rule. One approach is to apply ribbons analogous to ones Penrose invented for the kites and darts and impose the rule that two tiles can only be joined together if the ribbons match along the edge where they meet.

Another way to prevent an ordinary periodic pattern is to transform their straight edges into curves and notches analogous to puzzle interlocks, as shown in a beautiful example at the bottom of the facing page, which is made of individual wooden pieces. The tiling

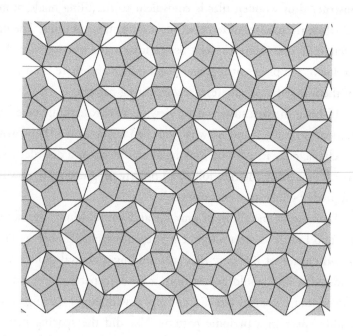

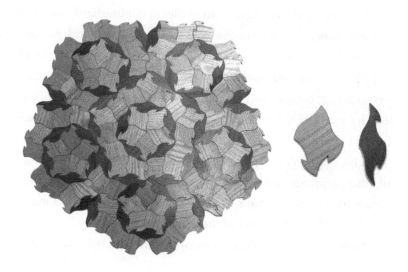

constructed of wooden tiles is equivalent to the tiling made of gray and white rhombuses in terms of the arrangement of units. The only difference is that interlocks have been added to the wooden tiles. With interlocks, the pieces fit together like a puzzle and there is no way to put them together in any regularly repeating pattern.

If this is the first time you are viewing a Penrose tiling, take a few moments to study it and see what your first impression is. How would you characterize it? Are you looking at an orderly or a disorderly pattern? If you think the tiles follow an orderly progression, how would you predict which tiles occur next?

Looking at the tiling of gray fat and white skinny rhombuses, Dov and I noticed certain frequently repeating motifs, such as a star-shaped cluster of five gray tiles surrounding a central point, which is not what one would expect from a random pattern. But we also noticed that those clusters did not repeat with equal spacing, as one would expect in a periodic pattern. Nor did the spacing between repeats appear to be arbitrary, as one would expect for a random pattern.

Comparing the configurations of tiles that immediately surround the star-shaped clusters, we observed that not all stars had the same surroundings. And when we considered the next layer of surrounding tiles, we observed even more differences. Study the figure on the previous page and you will note those differences. In fact, no two stars ever have exactly the same surroundings if one continues far enough out from the center of the star.

This was significant, because Dov and I knew it was the opposite of what one finds in a periodic pattern. Every tile in a square tiling always has exactly the same surrounding as any other tile no matter how far out from the center of the design you look.

From that simple observation, we confirmed that the Penrose

pattern could not be periodic. Then again, a pattern composed of clusters that were nearly the same and repeated frequently in the tiling could not be considered random, either. That led to the question: What kind of pattern could be both non-periodic and non-random at the same time?

There was no ready answer, which really intrigued me. No one had ever seen anything quite like a Penrose pattern before he invented them in 1974. Even Penrose himself did not appear to fully appreciate what he had invented. In his original paper, Penrose described his pattern as "non-periodic," which precisely defines what the tiling is *not*. It is not periodic. But it does not say what the tiling actually *is*. And that was the crucial issue for Dov and me.

The moment we began to study Penrose's tiling, we imagined that we might be able to construct an analogous three-dimensional pattern using a pair of building blocks. Then, by replacing each building block shape with a certain type of atom or cluster of atoms, we hoped to construct an atomic structure that would achieve our dream of a new type of matter.

But first, in order to show that the new atomic structure was truly novel and to figure out its distinctive physical properties, we needed to identify its symmetries. Merely describing the matter as non-periodic or non-random was not going to be good enough. So we spent the next few months focusing on Penrose's tiling to see if we could discover the mathematical secret to its symmetries.

The first remarkable property of Penrose tilings that Dov and I established was that they had a subtle kind of five-fold rotational symmetry, which was, of course, supposed to be impossible.

To see the five-fold symmetry of the Penrose tiling requires some effort. The image on the following page shows again an enlargement of the tiling composed of gray fat and white skinny rhombus tiles.

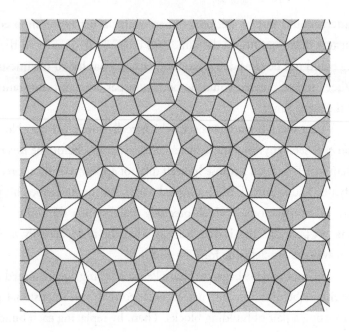

Take a moment and study the tiles that immediately surround any one of the star clusters. The arrangement is highly complex. Imagine rotating it one-fifth of the way around a circle, or 72 degrees. Does the arrangement look the same as the original?

If you try the experiment, you will find that the answer is "it depends." For some stars, the answer is "no." So ignore them and choose another. Continue until you find a star cluster for which the answer is "yes." You won't have to look very far to find one.

Now consider the second layer of tiles surrounding the star cluster you have chosen. Repeat the rotation by 72 degrees, one-fifth of the way around a circle, and ask if this larger configuration of tiles, which extends to two layers of tiles around the original star cluster, looks identical to the original.

Once again, the answer for some stars will be "no." So ignore those, too, and continue until you have found one of the rarer star clusters for which the answer is "yes." Now, repeat the process again for this subset going out three layers. And so on.

As you check more and more layers, you will be discarding more and more star clusters, but you will also find that there will always be *some* clusters remaining that have five-fold symmetry. The procedure is much more tedious than what is needed in order to check the symmetry for a periodic tiling, but it is still enough to prove that Penrose tiling has five-fold rotational symmetry.

A more sophisticated mathematical analysis can be used to show that, technically speaking, the Penrose tiling has more than five-fold symmetry. It actually has ten-fold symmetry. But for Dov and me, it made little difference whether the tiling had five- or ten-fold symmetry. Either way, the symmetry was strictly forbidden according to the mathematics of tilings and the established laws of crystallography.

That could only mean there was a faulty assumption underlying those laws that everyone had been missing for more than two hundred years. There was some kind of loophole. Once we realized that, Dov and I were hooked. We just had to find the loophole.

We already knew about matching rules, the mysterious interlocks that prevented the tiles from being put together in any kind of periodic pattern. Matching rules meant the shapes would only be allowed to fit together in patterns with the forbidden five-fold symmetry.

Using our ball and stick models, Dov and I had already begun to construct analogous three-dimensional structures composed of building blocks, where each block represented one or more atoms. For our model, we translated the Penrose interlocks into atomic bonds, which connected the atoms represented by one of our three-dimensional building blocks with those of another. The atoms would then be naturally prevented from solidifying into any type of crystal with a regular periodic pattern. The atoms would, instead, be forced to make the new type of matter with the icosahedral symmetry we were seeking.

This line of thinking was personally intriguing for me, because it was remarkably reminiscent of Vonnegut's imaginary ice-nine, in which a new arrangement of water molecules—ice-nine—was more stable than ordinary crystalline ice. The new form of matter we were pursuing, if we ever found it, might turn out to be a remarkably stable material that was harder than ordinary crystals. But what kind of regularity were the matching rules imposing?

One clue was that the Penrose tilings obey something called a "deflation rule." Namely, each fat and skinny rhombus in a Penrose tiling can be subdivided into smaller pieces that create another Penrose tiling. In the figures seen below, the original tiles have solid lines. The subdivision, or deflation, rule for each fat and skinny tile is indicated by the dotted lines. As shown on the right, the dotted lines join to form a new Penrose tiling with more pieces than the original.

Beginning with a small cluster of tiles, repeated deflation can produce a Penrose tiling with as many pieces as you wish. The inverse process, replacing a group of smaller tiles with larger tiles, is called an

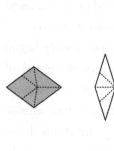

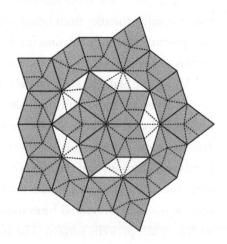

inflation rule. The deflation and inflation rules proved to Dov and me that the Penrose tiling has some kind of predictable hierarchical structure.

Dov and I were convinced that the combination of five-fold symmetry, matching rules, and deflation-inflation rules were unmistakable evidence that Penrose's arrangement of tiles was ordered in some novel, nonintuitive way, but exactly what kind of ordering was it?

It was all incredibly frustrating. Dov and I knew if we could answer that question, we would discover a loophole in the long-accepted rules about symmetry, which dictate what types of matter are possible. And that would be the key to a major paradigm shift and to discovering a range of materials unlike any seen before.

But what in the world could that loophole possibly be? We were stuck.

THREE

FINDING THE LOOPHOLE

PHILADELPHIA, 1982–83: Dov and I discovered an important clue to unlocking the secret symmetry of the Penrose tilings in the unpublished work of a brilliant amateur mathematician named Robert Ammann.

He was an unusual, reclusive man. Ammann was talented enough to be accepted at Brandeis University in the mid-1960s. But he only attended college for three years, during which time he rarely left his room. The administration finally dismissed him, and he never completed a formal degree.

Next Ammann studied computer programming on his own and obtained a job as a low-level computer programmer. Unfortunately, his job was eliminated during a company-wide cutback. So he began sorting mail at the post office, which was the kind of job that did not require much human interaction. His co-workers considered him uncommunicative and extremely introverted.

What the other postal workers probably never knew was that Ammann was a mathematical genius. Privately, he was engaged in the same world of recreational mathematics as the academic superstars Roger Penrose and John Conway. With typical modesty, Ammann simply described himself as an "amateur doodler with math background."

Dov and I stumbled across Ammann's ideas in two short papers in lesser-known journals, written by Alan Mackay, a crystallographer and professor of materials science at the University of London. Mackay shared our fascination with the icosahedron, Penrose tilings, and the fantasy of materials with forbidden five-fold symmetry. His two papers, which were more like speculative essays than research papers, presented some of his notional thoughts on the issue. They included two illustrations that piqued our interest.

In the first, Mackay showed a pair of fat and skinny rhombohedrons, illustrated below. The three-dimensional shapes were already very familiar to Dov and me. We knew they were the obvious three-dimensional analogs of the fat and skinny rhombuses that can be used to create two-dimensional Penrose tilings. So Mackay appeared to be on the same track that we were.

But we were disappointed to find that his paper did not present any *matching rules* that would prevent the three-dimensional building blocks from forming periodic crystal structures. It was essential for Dov and me to find those particular matching rules. Without them, the atoms would still be able to arrange themselves into any one of a number of ordinary crystal structures, instead of being forced into the impossible structure we were hoping to discover.

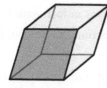

Fat Rhombohedron

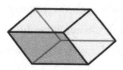

Skinny Rhombohedron

The second figure Mackay included (not pictured here) also intrigued us. It was a photo of a diffraction pattern created by shining a laser through the image of a Penrose tiling. From Mackay's image, it was clear that the complex diffraction pattern included some fairly sharp spots, including some at the corners of a decagon and some at the corners of a pentagon. But we could not determine if the spots were truly sharp or somewhat fuzzy pinpoints, or whether they were arranged along perfectly straight lines.

For physicists like Dov and me, those details were critically important. Truly sharp pinpoints aligning in perfectly straight rows, along with arrays of spots in the shapes of perfect decagons and pentagons, would be a diffraction pattern no one had ever seen before. That, of course, would indicate an atomic order no one had ever seen before.

Fuzzy spots with imperfect alignment would be much less exciting. That would indicate a combination of both atomic order and disorder, similar to the arrangements David Nelson and I had already studied, and not a new form of matter.

Clearly, the first possibility, which would indicate something truly novel, was what Dov and I were hoping for. But once we contacted Mackay to ask about matching rules and the precise mathematical nature of the diffraction pattern in his photo, he had no answers to our questions. Mathematics was not his forte, Mackay explained. So he did not know how to prove whether the diffraction spots for a Penrose tiling were perfectly sharp or somewhat fuzzy. He also confessed that he only had one photograph, which was unfortunate because a photograph always introduces a little distortion. So he could not be sure about the diffraction properties.

Mackay also informed us that the fat and skinny rhombohedrons he discussed in his paper were not his own creation. They were taken directly from the work of an unknown amateur named Robert Ammann. That was the first time we had ever heard mention of the mysterious genius who communicated with very few people other than Martin Gardner, the *Scientific American* guru of recreational mathematicians. Mackay suggested we contact Gardner for help.

Dov immediately wrote to Gardner, who in turn referred us to Branko Grunbaum and Geoffrey Shephard, who were writing an upcoming book about tilings that included some of Ammann's ingenious inventions. From them, we discovered that Ammann had independently invented rhombus tiles with matching rules that force five-fold symmetry similar to Penrose's discovery. Incredibly enough, he had also invented another set of tiles with matching rules that force the equally impossible eight-fold symmetry.

Ammann was not a trained mathematician, so he did not provide any proof that his matching rules worked and he never wrote a scientific paper to that effect. He just intuitively knew it to be so.

Gardner also provided us with some of Ammann's notes that expounded on his ideas for building blocks with icosahedral symmetry. But once again, there were no rigorous proofs nor any attempts at plausible arguments.

Several years later, Dov and I managed to track down the elusive genius in the Boston area and were able to entice him to visit us in Philadelphia. Ammann was every bit as brilliant as I envisioned. He was full of creative geometric ideas and intriguing conjectures that were never published, but which often turned out to be correct. Some, like his ideas about the rhombohedrons that first appeared in Mackay's illustration, were things that Dov and I had discovered independently as a result of hard work and painstaking proofs. For Ammann, everything was just intuitively obvious. Sadly, Ammann died several years later and Dov and I were never able to meet with him again.

Ammann's most impactful invention, as far as Dov and I were concerned, was his introduction of the eponymous Ammann bars, which were a powerful matching rule. Using rhombuses with perfectly straight edges, Ammann drew a set of bars across each fat and skinny rhombus according to the precise prescription shown below as dash-line segments.

Ammann's matching rule is that two tiles can only join together if a bar from each continues straight across any edge where two tiles meet. That produces the same kind of constraints as we had seen with Penrose's ribbons or interlocks. So at first blush, it was nothing remarkable.

But upon closer inspection, the Ammann bars changed everything. Dov and I discovered that the bars revealed something about Penrose tilings that even Penrose himself had not recognized. And *that* was the discovery that launched Dov and me into the strange new world of impossible symmetries.

Dov and I observed that when the tiles are joined together according to the matching rules, the individual Ammann bars connect to form Ammann *lines* that stretch across the entire tiling in straight lines. The image opposite shows the tiles and, superimposed, the

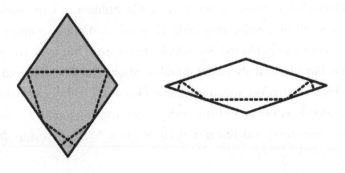

crisscrossing array of straight Ammann lines. The array consists of five sets of parallel lines oriented along different directions.

Dov and I found that each of the five sets of parallel lines are identical and that the angles between pairs of crisscrossing sets of lines were precisely the same angles as between the edges of a pentagon. This was the simplest proof we could ever imagine that the tiling had perfect five-fold symmetry.

For Dov and me, it was an absolutely thrilling moment. Now we knew for sure that we were heading for a discovery that would be in direct conflict with the centuries-old theorems of Haüy and Bravais. We were confident the Ammann lines held the clue to evading those established theorems, and to explaining the secret symmetry of Penrose tilings. But we still had to decipher their meaning.

The key was to focus on just one of the five sets of parallel lines, such as the set shown with thick lines in the image on the next page.

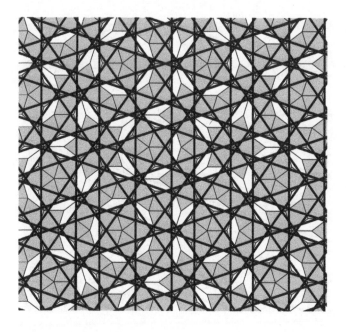

Here we see that the channels between parallel Ammann lines are limited to one of two possible widths, wide (W) and narrow (N). For us, the two most important things would be the ratios between the two channel widths and the frequency with which they repeated in the pattern. We were about to discover that those two features, the ratio and the sequence, were related to two very famous mathematical concepts called the "golden ratio" and the "Fibonacci sequence."

The golden ratio is often discovered in nature, and has been incorporated in artistic works since ancient times. The Egyptians are thought to have used it to design the Great Pyramids. In the fifth century BCE, the Greek sculptor and mathematician Phidias is purported to have used the golden ratio to create the Parthenon in Athens, which is considered a monument to Greek civilization. The ratio is sometimes signified by the Greek letter Φ, pronounced "phi," in honor of Phidias.

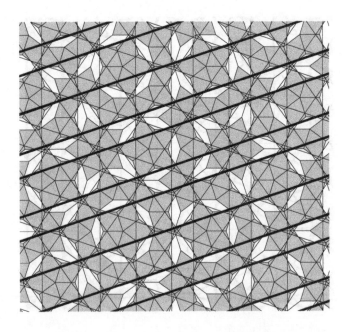

Euclid, the Greek mathematician who is considered the father of geometry, provided the earliest recorded definition of the golden ratio using a simple object. He considered how to break a stick into two pieces such that the ratio of the shorter length to the longer length was the same as the ratio of the longer length to the sum of the shorter and longer lengths. The solution he found was that the longer piece must be exactly Φ times the length of the shorter piece, where Φ is

$$\frac{1+\sqrt{5}}{2} = 1.61803\ 39887\ 49894\ 84820 \ldots$$

. . . a never-ending, never-repeating decimal.

Numbers with nonrepeating decimal forms are called irrational because they cannot be expressed as a ratio of integers. This is to be contrasted with rational numbers, like 2/3 or 143/548, which are ratios of integers and whose decimal forms, 0.333 and 0.26094890510948905109, are seen to regularly repeat if you follow the digits out far enough.

It was not all that surprising to Dov and me that the golden ratio is found in the five-fold symmetry of a Penrose tiling, because the ratio itself is directly related to the geometry of a pentagon. For example, in the left image below, the ratio of the length of the upper line that connects opposite corners of the pentagon to the length of one of the edges

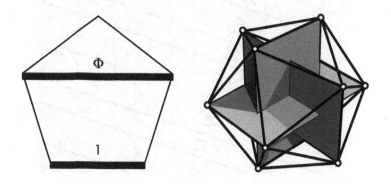

of the pentagon is golden. The icosahedron (pictured on the previous page, to the right) also incorporates the golden ratio; the twelve corners form three perpendicular rectangles, and each rectangle has a length to width ratio equal to the golden ratio.

What was surprising to Dov and me, though, was to discover that the golden ratio was incorporated in the sequence of (W)ide and (N)arrow channels, as well.

Consider the channel sequence of Ws and Ns in the figure below. It never settles into a regularly repeating rhythm. If you were to count the Ws and Ns, and then compute the ratio at certain points along the way, you would find that the ratio after the first three channels is 2 to 1; after the first five channels, 3 to 2; after the first eight channels, 5 to 3; etc.

There is a bit of simple arithmetic that can generate this sequence. Consider the first ratio, 2 to 1. Add the two numbers (2 + 1 = 3), and

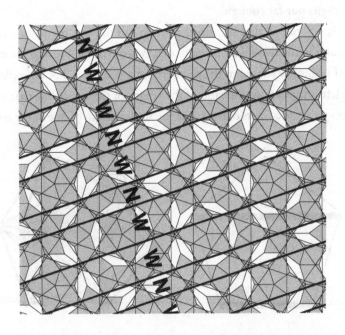

then compare the sum (3) to the larger of the original two numbers (2). This new ratio is 3 to 2, which is also the next ratio in the sequence of channels. Add the next two numbers (3 + 2 = 5), and once again, compare that number to the larger of the previous two: 5 to 3.

You could continue the process indefinitely to obtain 8 to 5, 13 to 8, 21 to 13, 34 to 21, 55 to 34, and so on. The ratios would precisely predict the sequence of Ammann channels.

Dov and I immediately recognized that sequence of integers: 1, 2, 3, 5, 8, 13, 21, 34, 55, . . . They are known as Fibonacci numbers, named after the Italian mathematician Leonardo Fibonacci, who lived in Pisa during the thirteenth century.

The ratios of consecutive Fibonacci numbers—2:1, 3:2, 5:3, . . . —are ratios of integers and, therefore, all rational. But a famous property of the Fibonacci numbers is that, as the integers get larger, the ratios get closer and closer to the golden ratio. That is how Fibonacci numbers and the golden ratio are interrelated.

And it turns out the only way to obtain a pattern of Ws and Ns that reproduces the Fibonacci numbers is to have the Ws repeat with a greater frequency than the Ns, as the Penrose tiling extends out in all directions, by a factor precisely equal to the golden ratio, an irrational number. And *that* is the secret of the Penrose tiling in a nutshell.

A sequence composed of two elements that repeat at different frequencies, the ratio of which is an irrational number, is called *quasiperiodic*. A quasiperiodic sequence never repeats exactly.

For example, no two channels in the Fibonacci sequence are surrounded by the same exact pattern of Ws and Ns, though one has to go out quite far in some cases to find the differences. The same applies to Penrose tiles. Check far enough out and you find that no two tiles have the exact same surrounding configuration.

At last, Dov and I could point precisely to the loophole in the centuries-old rules of Haüy and Bravais. The fundamental theorem of crystallography states: If a pattern of tiles or atoms is *periodic*, occurring with a *single* repeating frequency, then only certain symmetries are possible. In particular, five-fold symmetry along any direction is truly impossible for an arrangement of atoms that is periodic. We might call this the first kind of impossible, meaning absolutely inviolable, just like 1 + 1 can never equal 3.

However, when scientists asserted to generations of students that five-fold symmetry was impossible for matter of any type, it was a case of the second kind of impossible—a claim resting on an assumption that may not always be valid. In this case, physicists and materials scientists were assuming without proof that all orderly arrangement of atoms are periodic.

The Penrose tiling, Dov and I now understood, is a geometric example of an orderly arrangement that is not periodic. It is *quasiperiodic*, precisely described in terms of tiles or atoms with *two* different repeating frequencies with an irrational ratio. That was the loophole we had been seeking. Scientists had been assuming that atoms were always arranged either periodically or randomly in matter. They had not considered quasiperiodic arrangements before.

If real atoms could somehow be arranged into a pattern in which they repeated with two different frequencies whose ratio is irrational, the result would be a whole new form of matter that shattered the rules established by Haüy and Bravais.

It all seemed so simple, and yet so profound. It was as if a new window had magically appeared in front of us, a window that only Dov and I could peer through.

In the distance, I knew, was an entire field of potential breakthroughs. For now, that field was ours and ours alone to explore.

FOUR

A TALE OF TWO LABORATORIES

Dov and I did not realize it, but we had just entered a race against time. Ever since we had discovered that quasiperiodic order was the secret to creating matter with forbidden symmetries, we had been developing our theory about a new form of matter on our own timetable.

We had no concern that another theoretical physicist would duplicate our work. The peculiar approach we had taken, using recreational mathematics and tilings for inspiration, was far too unconventional to be mimicked. We had not published our ideas yet, so no one else could run ahead with them. And how could an experimentalist who never heard about our quasicrystal theory ever compete? That seemed impossible.

The one thing we did not anticipate was serendipity. Sometimes a simple experiment can produce an unintended outcome. And if the right person is paying attention, there is always the chance of a scientific breakthrough. As it turned out, during the same period of time that Dov and I were systematically developing our radical theory, an unknown scientist named Dan Shechtman, who was working in a lab less than 150 miles from us, had stumbled upon a seemingly nonsensical experimental result.

It was a bizarre coincidence that would become an unusual

footnote in the history of science. Two teams with no knowledge of the other were independently challenging the same rigid principles accepted for centuries. It would be two years before either team learned of the other. And once they did, it would quickly become apparent that each team needed the other to reach their goal.

PHILADELPHIA, 1983–84: Dov and I were meeting almost every day to flesh out our theory. Our immediate focus was to find a way to exploit the loophole we had discovered in the rules of crystallography: quasiperiodic ordering. Our aim was to use that loophole to create a three-dimensional structure with the forbidden symmetry of an icosahedron. It was an ambitious goal, but if we could show that such a geometrical structure existed, we could then begin to imagine how real-life atoms and molecules might be arranged the same way.

It sounded crazy. But it was the same idea that had motivated me from the beginning, first as a teenager inspired by Vonnegut's fictional ice-nine, and then, years later as a researcher, when a tantalizing hint of forbidden symmetry popped up in my work with David Nelson on rapidly cooled liquids.

Roger Penrose's discovery of how to design shapes with special interlocks to create his intricate patterns was a major achievement. For us to repeat that feat in three dimensions would be, in many ways, much more challenging.

The icosahedron, like every other three-dimensional object, has different rotational symmetries in different directions. The forbidden five-fold symmetry appears along *six* different directions. Two- and three-fold symmetry can be seen if the icosahedron is viewed along other directions.

Dov and I began by working with rhombohedrons, the three-dimensional equivalent of the rhombuses Penrose had used for his flat designs. We knew that rhombohedrons can be packed together in a *periodic* arrangement, as Haüy first discovered more than 200 years ago in his explorations of calcareous spar. But Penrose had found matching interlocks for his rhombuses that could prevent any such periodic patterns. The interlocks forced his fat and skinny rhombuses into a quasiperiodic arrangement. We needed to prove whether the same sort of mechanism existed for fat and skinny rhombohedrons. Dov and I discovered that we needed to use twice as many units as Penrose—two fat rhombohedrons and two skinny rhombohedrons, each with unique interlocks. More shapes, more interlocks, more complications.

As always, we found it useful to construct physical models of the abstract theoretical objects we were studying in order to visualize the structure. So once again, we turned my office into a comical-looking arts-and-crafts studio.

The easiest part was creating the two types of building blocks. We designed cardboard cut-outs of the fat and skinny rhombohedrons that could be folded to form the four different units—two fat and two skinny shapes. We tried taping them together according to our proposed interlock rules, but the process turned into a sticky nightmare. So we rolled up our sleeves and glued magnets into the corners of all the cardboard cut-outs. The magnets were precisely placed so they could play the same role as interlocks. The blocks would only stick together if the three-dimensional interlock rules were satisfied. It was highly organized chaos, or so I kept telling befuddled visitors to my office.

On the next page are photographic examples of some of our constructions. On the top left there is one cluster of ten fat and ten skinny rhombohedrons joined together in a nearly spherical shape.

The outside surface of this cluster has a daunting name. It is a "rhombic triacontahedron," which is Greek for "thirty faces on the surface, each with the shape of a rhombus."

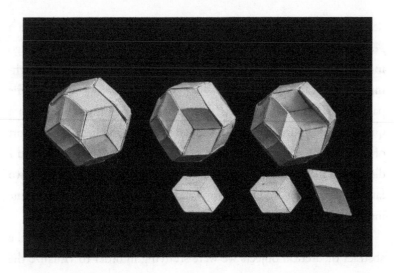

In the middle image, a skinny rhombohedron has been removed, revealing a bit of the interior. On the right, a fat rhombohedron has also been removed to reveal even more.

The rhombic triacontahedron was the first step, we showed, in packing our fat and skinny rhombohedrons together in a quasiperiodic pattern of arbitrarily large size, while maintaining the symmetry of an icosahedron. Equally important, there were no gaps between our building blocks, the rhombohedrons, and our new interlocks specifically prohibited them from forming any other kind of structure, including the regular periodic arrangements of a crystal.

Now that we knew a three-dimensional quasicrystal was theoretically possible, we needed to identify groups of atoms that could join together in analogous ways, with analogous matching rules, so that a quasicrystal would be the only possible result.

We began considering what other previously forbidden rotational symmetries were possible with quasiperiodic order. The answer, incredibly enough, was all of them. Seven-fold, eight-fold, nine-fold . . . literally an infinite number of new possibilities once thought to be forbidden were now allowed. A beautiful example of a quasiperiodic tiling with seven-fold symmetry is shown below.

Dov and I were now rapidly making so many discoveries and had so many new directions to explore that it was hard to decide when to stop researching and start writing a scientific paper. Not believing that there were any other competitors on the field, which might have given me a reason to rush, I made the fateful decision to continue working and delay publishing our results until we had made further progress.

The early 1980s was one of the most fertile times of my career. Dov was not the only talented graduate student I was working with. Andy Albrecht and I were focused on an exciting new idea in cosmology, the inflationary theory of the universe, which had just been introduced by a physicist at the Massachusetts Institute of Technology named Alan Guth.

Few scientific theories are ever complete when first introduced, and

the inflationary theory was no different. Alan had proposed that inflation, a hypothetical period of rapid expansion a few instants after the big
bang, could potentially explain, in part, why the distribution of matter
and energy in the universe is so uniform today. In order to do that,
however, he had to assume that inflation would stop after a short period
of time. But there was the rub. Alan could not find any way to make
inflation stop. Andy and I, along with Andrei Linde, a theorist working
independently in the Soviet Union, solved that critical problem.

Our "new inflationary theory" took hold rapidly. It had an explosive effect that triggered a period of prolific innovation in cosmology, astrophysics, and particle physics that continues today. Unlike
my work with Dov on new forms of matter, the exploration of new
inflationary cosmology was a crowded field with many sharp-elbowed
competitors. There were many important follow-up projects that simply could not be ignored.

Throughout this time, though, I was also quietly testing reactions
to our new quasicrystal theory. I had begun discussing it informally
with well-known condensed matter physicists and materials scientists.
But to my surprise, the response was uniformly discouraging:

*You and Dov have an imaginative idea for a new form of matter
that might be mathematically possible, but it seems far too complicated,
compared to the simple case of a periodic crystal, to exist in the real world.*

I could understand their attitude. After all, Dov and I were challenging centuries of scientific wisdom by proposing a new state of
matter based solely on the study of abstract tilings. What was needed
was experimental proof that there existed combinations of atoms that
would arrange themselves into a true quasicrystal. Without that, our
idea was just another theoretical fantasy with no connection to reality.

I was more sensitive to the criticism than Dov, who wanted to
publish our basic idea right away. I wanted to wait until we had developed a more concrete proposal. I also wanted to be able to make a

testable prediction, a necessary component of any scientific theory, to explain how to identify the new form of matter through experiment. Without that, I concluded, our work would probably be discounted. So there was no point in publishing.

In 1983, Dov and I reached a compromise. We agreed to protect our intellectual investment by submitting a patent disclosure, which we filed with the University of Pennsylvania's Technology Licensing office. The submission would present our concept and formally establish priority. But we would not release our idea to the general scientific community until we had made more progress.

The disclosure, partially reprinted below, described our building blocks, the rhombohedrons, and the matching interlocks. It explained that the connections were designed to force the building blocks into a noncrystalline pattern with the symmetry of an icosahedron. It also

---

UNIVERSITY OF PENNSYLVANIA
Philadelphia, Pennsylvania 19104
Instructions: See Reverse Side. PRINT OR TYPE all information

INVENTION DISCLOSURE
No. UP   -

Inventor(s) Full Name	Office Address & Extension	Home Address	Citizenship
Paul Joseph Steinhardt	2N9D, David Rittenhouse Lab, El X5949	109 Valley Forge Terrace, Wayne, Pa. 19087	USA
Dov Irving Levine	2W1N, David Rittenhouse Lab, El X6214	919 Lombard St. Phila., Pa.	USA

Title of Invention (short & descriptive):

CRYSTALLOIDS

Description of Invention: (if more space is needed, use plain white paper. Sign, date, and have each sheet witnessed.)

The crystalloid was invented as a result of a recent investigation by D. Nelson, M. Ronchetti and one of us (PJS).[1] Our computer simulation studies indicate that the bonds that join atoms in simple supercooled liquids and glasses are, on average, oriented along the axes of an icosahedron, even though the bonds are randomly spaced. The crystalloid was invented by Dov Levine and Paul Steinhardt as an idealization of such a structure. A real material with atoms placed at the vertices of a crystalloid would represent a new phase of matter with properties different from either liquids or crystals. We are continuing to study the physical properties of such a new phase and plan to publish our findings in a journal article and in Dov Levine's thesis.

Inventor Signature (date)	Inventor Signature (date)	Inventor Signature (date)

Disclosed to and understood by:

A.F. GARITO

Witness Signature (date)	Witness Signature (date)	Witness Signature (date)
A.F. 6—. ⊃ 4-23-83		

0176   WANG

explained how the idea could potentially lead to a new phase of matter with properties different from either liquids or crystals. Dov and I called our theoretical invention "crystalloids" in the 1983 disclosure, but later renamed them "quasicrystals."

Was this just an abstract theory, as critics claimed, or was it actually a valid scientific theory and somehow testable? And how would we ever recognize a quasicrystal if we were lucky enough to find one? Dov and I spent months grinding away at calculations only to find that the answer was relatively simple. An ordinary X-ray or electron diffraction pattern would reveal the quasiperiodicity and forbidden symmetry of its atomic arrangement.

Compared to a crystal, a quasicrystal's diffraction pattern has a much richer, more complex structure partly because it is composed of atoms that repeat with different frequencies related by an irrational number, such as the golden ratio.

If X-rays or electrons magically diffracted from only one type of atom in the quasicrystal, it would produce true pinpoint diffraction, known as Bragg peaks, with equal spacing between the peaks. But in reality, X-rays and electrons diffract from all of the atoms in a quasicrystal. Different subgroups have different pinpoint diffraction patterns as well as different spacings between the atoms. An icosahedron has multiple symmetries, which also adds to the complexity.

Our predicted diffraction pattern took different forms depending on whether the X-ray or electron beam was shone down an axis of five-fold, three-fold, or two-fold rotational symmetry. The image on the opposite page shows our computed prediction when the beam is aimed along the "impossible" five-fold symmetry.

We had deciphered the mathematical formula behind the secret symmetry and were able to make a bold quantitative prediction that could be tested by experiment: The diffraction pattern of

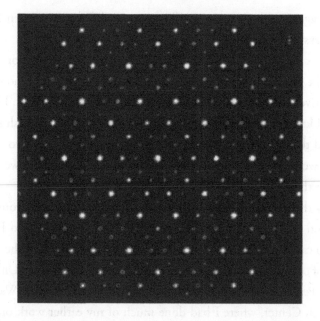

a quasicrystal would be composed of absolutely perfect pinpoints arranged in a snowflake pattern.

The archival figure shown above is the first such pattern ever computed. Our computer code drew a circle centered at each predicted pinpoint. The radius of each circle was chosen in proportion to the predicted intensity of diffracted X-rays. The figure we created was the first visual representation of the bright and dim spots that we would expect to see in the diffraction pattern of an actual quasicrystal.

If it were possible to see the ever-dimmer points, one would find that between every pair of spots there were yet more spots. And between every pair of those spots, there would be even dimmer spots, and so on. If Dov and I had created a circle for each predicted spot, the pattern would have been so crowded that the circles would have merged into a white featureless cloud. We knew that experiments would only be able to detect the brightest spots. So we figured that our image would approximate the signature diffraction pattern for a quasicrystal.

By creating the figure, Dov and I had made a prediction that could be used to test and potentially disprove our theory. So now, we had arrived at another milestone. *Time to publish?* Once again, I held back. I knew there was something else we needed to accomplish if we wanted the radical new theory to be taken seriously. We had to show that it would be possible to replace the rhombohedron blocks we used for our theoretical model with real matter.

By the summer of 1984, the demanding commitments from my work on the new inflationary theory had finally subsided. So I was able to carve out the large block of time needed to focus on the final stage of our quasicrystal research. I took a sabbatical from the University of Pennsylvania and headed back to the IBM Thomas J. Watson Research Center, where I had done much of my earlier work on the atomic structure of amorphous metals.

My plan was to work with crystal experts to try to create the world's first synthetic quasicrystal. But unbeknownst to me, someone else had already done that. It was much easier to do than I had imagined. In fact, the discovery was a complete accident.

GAITHERSBURG, MARYLAND, 1982–84: "No such animal!" Dan Shechtman reportedly thought, as he looked at the strange sample under his electron microscope. The forty-one-year-old Israeli scientist had accidentally come across a material with all of the impossible properties Dov and I had anticipated, although he had no inkling about any of our ideas or any understanding of what his discovery actually signified. But Shechtman recognized that he was looking at something remarkable. And it would ultimately earn him the 2011 Nobel Prize in Chemistry.

Shechtman was working as a visiting microscopist at the National Bureau of Standards with John Cahn, whom he had met as a graduate student at the Technion, Israel's premier technological institute. Cahn was considered a major figure in condensed matter physics and was especially famous for his work on the processes that occur when hot metallic liquids are cooled and solidified.

Cahn had invited Shechtman to take a two-year leave from the Technion to work on a large project being funded by the National Science Foundation and the Defense Advanced Research Projects Agency (DARPA). The goal of the project was to synthesize and categorize as many different aluminum alloys as possible by rapidly cooling liquid mixtures of aluminum and other metals. Other scientists would create the alloys. Shechtman would use the electron microscope to study, identify, and classify the samples. It was an important service to the materials community because aluminum alloys are useful in many applications. But it was also a relatively dull and tedious assignment.

Robert Schaefer, one of the metallurgists at the lab, was especially interested in creating alloys composed of aluminum and manganese because of its superior strength as compared to pure aluminum. He and his colleague Frank Biancaniello made a series of samples composed of aluminum combined with varying amounts of manganese, and each sample was dutifully sent to Shechtman for analysis.

On April 8, 1982, Shechtman studied a sample of rapidly quenched Al_6Mn (scientific shorthand for an alloy with six atoms of aluminum for every atom of manganese), which had tiny feathery grains with roughly pentagonal shapes. A larger sample with beautiful flowerlets and clearly evident five-fold symmetries was later synthesized by An-Pang Tsai and his team at Tohoku University, and is shown on the next page.

When Shechtman fired a beam of electrons through the grains to obtain its diffraction pattern, he found something shocking. The pattern had what initially appeared to be rather sharp spots, as expected for crystal. But to Shechtman's surprise, the spots revealed an apparent ten-fold symmetry, which he, as well as every other scientist in the world, knew was impossible.

Shechtman sketched the pattern on one side of a page in his notebook. On the other side he noted a partial catalog of the diffraction peaks in which he wrote "10-fold???"

When Shechtman showed his results to his colleagues they were not particularly impressed. They, too, had been taught that true ten-fold symmetry was impossible. Everyone assumed the strange diffraction pattern could be explained by something called "multiple twinning."

A crystal twin is commonly formed when two crystal grains oriented at different angles grow together. A multiple twin is when three or more grains oriented at different angles combine. Two examples are shown in the images on the opposite page. The one on the left is an example of "triple twinning." It is easy to see with the naked eye that the combined crystals are oriented at three different angles.

The image on the right is much more subtle. It is an example of multiple-twinned gold. The sample is composed of five distinct

 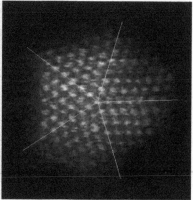

wedges, which have been made more obvious with added lines. The atoms are blurry white spots within each wedge. At first glance, the overall shape suggests a quasicrystal with five-fold symmetry. But that would be a false conclusion. This is not a quasicrystal.

Under the microscope, it becomes clear that each of the five wedges is made up of a regularly repeating hexagonal pattern of atoms. Therefore, each individual wedge is a crystal conforming to all the rules of crystallography. Taken as a whole, this is an example of a multiple-twinned crystal. It is a collection of crystals that just happened to come together in five wedge-shaped pieces forming the shape of a pentagon. Any solid composed of a combination of crystal wedges is always defined as a crystal no matter the number of wedges or how they are arranged.

Multiple-twinning is an everyday occurrence. So it was natural for Shechtman's colleagues, including John Cahn, to be convinced that the Al_6Mn sample was merely another example of the same phenomenon. No one was expecting to find anything the least bit unconventional in the midst of a dull survey of aluminum alloys. The lab dismissed the notion that Shechtman had discovered anything remarkable.

Shechtman, however, did not agree. He refused to relent and continued to press his case with the senior scientists. It was something novel, he argued. Unconvinced, John Cahn told him that there was a test that could settle the issue. Cahn told Shechtman to focus an electron beam on a very narrow region of the sample. If the sample was a multiple-twinned crystal, as the rest of the lab suspected, many of the spots in the ten-fold pattern would disappear and the remaining spots would form a pattern with one of the well-known crystal symmetries. On the other hand, if the sample truly violated the long-established principles and was uniformly ten-fold symmetric, all of the spots marking ten-fold symmetry would continue to appear no matter where the beam was focused.

Shechtman went back to his microscope and performed the crucial test. Wherever he looked in the Al_6Mn sample he found the same impossible ten-fold symmetry. It was an astonishing result that eliminated the routine explanation of multiple-twinning. History does not make clear, however, whether he showed the results to Cahn or anyone else in the lab before he completed his two-year term in America and returned to Israel.

What is known, however, is that Shechtman never gave up. He realized that his discovery was so outrageous that he would never be taken seriously without offering some plausible explanation. But he was an electron microscopist, not a mathematically-trained theorist. So he later teamed up with an Israeli materials scientist named Ilan Blech, whom he hoped could provide a possible theory.

With Shechtman's encouragement, Blech proposed a model based on a series of assumptions. First, he assumed that the aluminum and manganese atoms would somehow group together in identical icosahedral clusters. Then, he assumed the icosahedral clusters were thrown together in a random arrangement as the aluminum-manganese liquid cooled and solidified. He further assumed that all the clusters would somehow arrange themselves with the same orientation throughout the solid. That idea was equivalent to assuming one could randomly throw dozens of icosahedron-shape dice from the *Dungeons & Dragons* game into a bowl and have them land miraculously such that their points are all aligned in the same directions. The model was based on a stack of assumptions, some of which seemed unlikely to occur in real matter.

The figures below summarize the idea. The figure on the top shows a pair of adjoining icosahedrons with their points aligned. The figure on the bottom is a rough representation of how the random structure would appear.

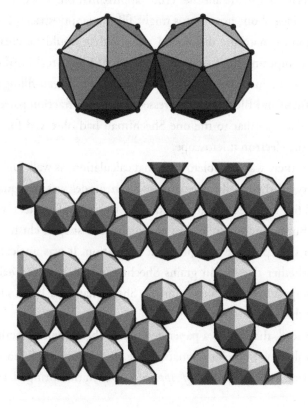

The drawing reveals the presence of large empty spaces between the icosahedrons when many are packed together according to Blech's ideas. Dov and I had encountered the same problem when we had tried to build clusters out of Styrofoam balls and pipe cleaners. We already knew that empty gaps were a big problem because they would not remain empty. There would be no way to stop atoms from moving in to fill gaps as a liquid cooled. And once they did so, the atoms would exert enormous pressure on the icosahedral clusters and would break up their delicate arrangement. That was one of the reasons Dov and I eventually abandoned the idea of using icosahedral clusters as building blocks. Our quasicrystal model uses rhombohedrons, which can pack together without any gaps.

Blech then made another crude approximation. Since he had no specific idea about how atoms might fill the empty spaces, he could only approximate the diffraction pattern that would be created by atoms composing the icosahedral clusters. Without real justification, he did not include the contribution from any of the gap-filling atoms. Shechtman and Blech were impressed that the diffraction pattern was qualitatively similar to the one Shechtman had observed for Al_6Mn under the electron microscope.

But there was a problem with that calculation, as well. Unlike our quasicrystal theory, the Shechtman-Blech model was not quasiperiodic. They assumed the arrangement of icosahedral clusters was random. But a random collection of icosahedron-shaped clusters could not possibly produce true pinpoint diffraction. It was unclear at the time whether the Al_6Mn grains Shechtman had observed exhibited true pinpoint diffraction or not. So Shechtman and Blech chose to ignore that issue.

Instead, they wrote a paper describing Shechtman's experimental results along with their explanation, the Shechtman-Blech model, and submitted it to the *Journal of Applied Physics* in the spring of 1984.

The paper was immediately rejected. The editor did not find either the experimental results or the theory compelling, and did not proceed with the next round of review, which would have involved circulating it among other scientists for comment.

Dov and I had not published anything yet. So Shechtman and Blech were completely unaware of our work. They had no idea that Dov and I had a fully developed theory that avoided all of the flaws of their model, or that our work could potentially explain the strange Al_6Mn sample. Conversely, since Shechtman and Blech's paper had been rejected before being circulated for peer review, Dov and I had no knowledge about the contents of Shechtman's laboratory notebook.

If there had ever been any exchange between our two teams there is a good chance we would have joined forces and presented the theory and experiment together.

But history unfolded somewhat differently.

FIVE

SOMETHING EXCITING
TO SHOW YOU

Most scientific breakthroughs are recognized slowly, like viewing a ship gradually emerging from a thick fog. But the discovery that quasicrystals are a reality, and not just a hypothetical idea, happened in a flash. I was fortunate enough to be there when it happened, and it was an unforgettable experience.

YORKTOWN HEIGHTS, NEW YORK, OCTOBER 10, 1984: It began as a fairly unremarkable autumn day. I was on a leave from the University of Pennsylvania and spending a few months at the IBM Thomas J. Watson Research Center, just north of New York City. I was hoping to work with other scientists at the lab to create the world's first synthetic quasicrystal.

Harvard physicist David Nelson, my former collaborator, was giving a seminar at the Center that afternoon and planned to stop by my office for a brief visit. Dov would be there too, because we planned to surprise David. I was anxious to share our work on the wild idea for a new form of matter that had grown out of the earlier work we had done together on rapidly cooled liquids.

David and I had not seen each other for several years and greeted each other warmly. He had the same clean-cut, boyish looks set off by wire-rim glasses that I remembered. I had been looking forward to the meeting because I knew he would appreciate what Dov and I had to show him.

Dov and I had applied for a patent about our idea the previous year, but we had not distributed the disclosure to other scientists. Lawyers at the University of Pennsylvania had recently concluded that, although our idea was "an important discovery . . . novel and unobvious . . . the utility of the discovery is still speculative." For similar reasons, we had not yet submitted our quasicrystal theory to a scientific journal. It was clear that we needed experimental proof to back up our "speculative" claims before we could publish our idea. So when David arrived and sat down to talk, he knew nothing about our work.

I began by telling him that Dov and I had something exciting to show him. But before I could say another word, David interrupted to say that he had something exciting to show *me*. We all laughed, and agreed the guest should be allowed to go first.

David reached into his briefcase and pulled out a "preprint," which is a typed version of a scientific paper submitted to a professional journal for rigorous peer review before being accepted for publication. It was common practice then, as now, to share and discuss preprints. But circulation methods were much less efficient pre-Internet.

The paper had been submitted by the team of Dan Shechtman, Ilan Blech, Denis Gratias, and John Cahn.

I was immediately stunned by the title: "Metallic Phase with Long-Range Orientational Order and No Translational Symmetry." *Wait a minute,* I thought. *No Translational Symmetry?* That suggested that the atoms in their material were randomly distributed. *Orientational Order?* That suggested the interatomic bonds were aligned.

The title, and the fact that David was showing it to me, led me to believe that the paper must be related to the computer simulations we had performed three years earlier while testing his "cubatic" idea.

That must be why he is showing me the preprint, I thought. *It appears to be experimental verification of our earlier findings.*

I quickly scanned the summary abstract to see if my first impression was correct and suddenly felt myself becoming alarmed. The scientists were studying a strange new alloy of aluminum and manganese when they found . . . *Oh my God!* . . . "sharp diffraction spots arranged with icosahedral symmetry."

I felt my pulse quicken. This was definitely not what David and I had been working on. This was more like the concept of quasiperiodic crystals that Dov and I had invented but not yet published.

Has this other team scooped us? I thought.

I raced through the rest of the abstract and was relieved to see the answer was no. There was no theoretical model included because, as I would later learn, the Shechtman-Blech model had been judged unconvincing. The paper was merely the announcement of experimental data, with no attempt at theoretical explanation. The years of work Dov and I had performed had not been duplicated.

With my competitive instincts safely assuaged, I began flipping through the rest of the preprint for more details. I caught my breath when I got to page 8, because I suddenly found myself staring at a very familiar diffraction pattern. It matched the pattern Dov and I had predicted for a quasicrystal, with the tell-tale symmetry of an icosahedron. *Impossible.*

I felt my chest starting to pound and fireworks going off in my head. I immediately recognized what this meant.

Quasicrystals exist! Here is proof that the crazy idea Dov and I were pursuing was actually not so crazy!

I knew it was a singular moment. For however brief a time, I was the only person who had seen both the experimental pattern and the

theoretical one. I was the only person on earth who knew for a fact that quasicrystals had just become a scientific reality.

I did my best to keep a blank expression on my face in order to preserve the moment and have a bit more time to savor the experience. After another moment, I jumped up from my chair, still without saying a word, and stepped across the office to pick up a single sheet of paper from my desk that I had prepared for the meeting. I continued to try to suppress the smile on my face as I slowly walked back to where David and Dov were sitting.

"Here, David," I said as calmly as possible, "is the exciting thing we wanted to show *you*."

In my right hand was the sheet I had just picked up on my desk, which showed the pattern of diffraction peaks that Dov and I had predicted for a quasicrystal. In my left hand, I held the preprint, turned to the page with the experimentally measured diffraction pattern.

The two patterns were a match.

Dov, already familiar with our work, reacted immediately. *"Oh my God!"*

I was not sure what David was thinking. But Dov and I had no doubt about what had just happened. Two scientific groups, working in laboratories only 150 miles apart and totally unaware of each other's efforts, had managed to make completely independent and yet totally complementary breakthroughs.

Dov and I had invented a theory of quasicrystals, but we had no experimental proof. Shechtman's paper had an experiment, but no theoretical explanation. Each group had separate pieces of the same puzzle. Together, it was a radical, and yet fundamental, discovery about nature.

David began asking questions about how we had predicted the snowflake diffraction image. Dov and I tried to answer his questions and explain our research in detail. But in truth, we could barely contain ourselves and were both sputtering with nearly uncontrollable excitement throughout the rest of the meeting.

Dov and I were elated because our theory could explain the seemingly impossible experimental result. But unfortunately, it also meant we had no time to celebrate. I told Dov that we had to drop everything else we were doing, and pull together all the results we had accumulated over the last three years. We needed to identify the most important points and write a short announcement paper for *Physics Review Letters*. Then, we needed to prepare a much longer paper with the full results.

I knew all of this could be accomplished quickly because we had already completed an enormous amount of research. It was just a matter of prioritizing the material and choosing which parts to present and in what order.

We began the *Letters* paper by introducing the concept of quasicrystals. We explained that they were new forms of matter with a quasiperiodic arrangement of atoms and a symmetry that was long thought to be impossible. We showed that solids with this property have electron diffraction patterns comprised entirely of sharp Bragg peaks. There were no fuzzy peaks, and no diffuse cloud connecting them. We explained our atomic building blocks, the rhombohedrons, along with the proposed matching rules we had invented that enabled the atoms to fit together in a quasiperiodic pattern. We also presented illustrations of the predicted diffraction patterns, the culmination of three years' worth of theoretical research.

Then, we turned our attention to the Shechtman team's results. Since their paper had not yet been reviewed or published, there was still a chance that their alloy might turn out to be something other than a quasicrystal. So Dov and I were conservative and did not claim an exact match:

> We show that the recently observed electron diffraction pattern of an aluminum-manganese alloy is closely related to that of an icosahedral quasicrystal.

Less than three weeks after our fateful meeting with David Nelson, Dov and I submitted our paper presenting the theoretical explanation for this bizarre new form of matter. We formally introduced its name to the scientific community in the title of our paper: "Quasicrystals: A New Class of Ordered Structures."

At this point, Dov and I were ready to reach out to the experimental group to tell them our exciting news. As it turned out, though, David Nelson had already written to John Cahn at the National Bureau of Standards to let him know that Dov and I had already developed a theory that might be relevant. So there was not much need for me to introduce myself. We quickly arranged for John and his colleague and coauthor, French crystallographer Denis Gratias, to visit Dov and me in Yorktown Heights.

John Cahn was a large man with a kindly face. We had never met before, but unbeknownst to him, I had a strong professional connection to what I considered to be some of his most important work. John began our meeting by explaining his background, particularly his work on a little-known process called "spinodal decomposition," a process that can occur during the solidification of metallic liquids.

John mentioned, almost as an aside, that he had heard there was a cosmologist who was using those ideas to develop a new theory of the early universe. I was a cosmologist. Did I know anything about that? he asked.

"Yes," I said, "there is indeed a certain cosmologist I know who is using your experimental findings to help develop his theories. In fact," I said with a smile, "that person just happens to be me." John's theory of spinodal decomposition was actually my key inspiration in developing a new inflationary theory of the universe, which introduced what is known as a "graceful exit" from the initial, explosive inflationary process. "It is an honor to finally meet you," I told him.

After a brief discussion about our cosmic connection, we settled down to business and spent the next five hours excitedly comparing notes about quasicrystals. Each team explained their parallel histories, one experimental and one theoretical, which, to everyone's amazement, had just crossed paths with fateful consequences.

John explained how his protégé, Dan Shechtman, had first discovered a ten-fold diffraction pattern in an alloy fabricated in 1982 at the National Bureau of Standards. When Dan had shown him the pattern, John prescribed a series of tests in order to exclude the most likely explanation, which was that the alloy was an ordinary multiply-twinned crystal.

John told us that he heard nothing more about the subject until two years later, in 1984. Dan returned to the lab with the results of the multiple-twinning test, along with a description of the model that he and Ilan Blech were proposing to explain the strange new alloy. Dan told him their paper had already been rejected by the *Journal of Applied Physics*.

John was greatly impressed by Dan's improved data, especially the results that showed that the multiple-twinning idea was invalid. He was less impressed, though, by the Shechtman-Blech model, which he considered sketchy and flawed.

So instead of pursuing that theory, John had recommended that Dan focus solely on reporting the experimental results. He suggested that a short paper be submitted to the prestigious *Physical Review Letters*. Dan accepted the advice and invited John to join him as a collaborator and to help rewrite the paper. John, in turn, contacted Denis Gratias, a French crystallographer, to join their team and double-check the analysis. The end result was the preprint David had brought me, submitted for publication by the team of Shechtman, Blech, Gratias, and Cahn.

John told us he was already trying to duplicate the inexplicable experimental results. His lab group had begun further studies to strengthen their conclusions about the unusual alloy and to search for other materials that might have similar diffraction patterns.

Then it was our turn. Dov and I described in great detail how we had arrived at our ideas and recounted all the research we had performed in the last three years. Most importantly, we showed them our predicted diffraction pattern for a quasicrystal with the symmetry of an icosahedron. All of us noted its close agreement with the measured diffraction pattern for the aluminum-manganese alloy reported in the preprint.

It was an intoxicating, exhausting, and exhilarating meeting.

A few weeks later, I made the first public presentation of the quasicrystal theory at a venue that held particular significance for me. I spoke at a specially arranged seminar at the Laboratory for Research on the Structure of Matter at the University of Pennsylvania. The lecture hall was packed. It was a homecoming of sorts for me, and our results were received enthusiastically. I was enormously grateful to the leaders and members of the Laboratory because they had steadfastly provided both encouragement and financial support throughout the prior three years, even when our quasicrystal idea seemed to have questionable scientific value.

John Cahn did me the honor of traveling more than two hours from Gaithersburg, Maryland, to attend the lecture. Once I had finished my presentation, he gave me yet another great honor by standing up and publicly endorsing our theory. John announced that in his view, our quasicrystal model correctly explained their team's new material.

With our paper submitted and the first public presentation completed, I finally took some time to reflect on what we had just accomplished. The scientific fantasy I had secretly nursed since high school and floated practically on a whim in a university lecture was a scientific reality. I was struck by what seemed to be a logical extension of that new reality:

If quasicrystals are a fundamental new form of matter that truly exist, as the laboratory evidence shows, then surely they must exist in nature!

Perhaps they are hiding right under our noses, I thought. We just have to figure out where to look for them. There might even be quasicrystals on display right now in museums that have been misidentified as crystals.

The thought got me very excited. Over the next few months, I made targeted inspections of the mineral collections at several museums, including the Franklin Institute in Philadelphia, the American Museum of Natural History in New York, and the Smithsonian National Museum of Natural History in Washington, D.C. I went from display case to display case hunting for a misidentified quasi-crystal. It was such a wild hunch that I never tried to talk to anyone at the museums and eventually came up empty. Perhaps my insight about the possibility of natural quasicrystals was not so insightful after all.

The Shechtman team's paper announcing their experimental results appeared in *Physical Review Letters* on November 12. Our theoretical explanation of those results appeared in the same journal on December 24, the penultimate issue of 1984.

Perfect timing and a perfect fit, I thought.

The two papers drew attention and strongly positive responses from scientists and journalists from around the world. There were articles in scientific journals as well as the consumer press, including *Physics Today*, *Nature*, *New Scientist*, and the *New York Times*. The *Times* article, headlined "Theory of New Matter Proposed," described how we had "postulated a new quasicrystalline state of matter that apparently explains the bewildering results of a test recently conducted at the National Bureau of Standards."

As news of the breakthrough spread across the globe, Dov and I were surprised to learn about scientists in other parts of the world who had been developing related ideas. Some were interested in the mathematics of Penrose tilings; some were interested in quasiperiodicity; some were even thinking about materials with icosahedral symmetry. In the pre-Internet days, it was much more difficult to share information. So Dov and I were previously unaware of these papers because they were not published in journals well known to physicists. But now, we were being contacted by the authors and, in turn, devouring everything they had written.

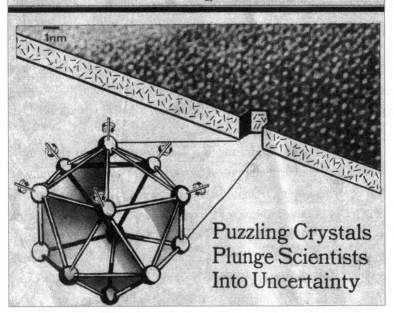

Theory of New Matter Proposed

TUESDAY, JULY 30, 1985

Copyright © 1985 The New York Times

Science Times

The New York Times

1nm

Puzzling Crystals Plunge Scientists Into Uncertainty

We were particularly struck by the work of Dutch mathematician Nicolaas de Bruijn, who had written a series of beautiful papers in 1981 with an ingenious "multigrid" method for generating two-dimensional Penrose tiling patterns without relying on any of the normal matching or subdivision rules. Dov and I teamed up with another talented young graduate student at Penn named Joshua Socolar to further develop those ideas. The three of us were able to generalize De Bruijn's multigrid method to create quasiperiodic patterns with *any* symmetry in *any* number of dimensions including purely mathematical constructs beyond three dimensions.

Our generalized multigrid method demonstrated in a straightforward, direct way something that Dov and I had already proven in a more abstract, indirect mathematical way: Quasicrystal patterns could be made for an *infinite* number of different symmetries forbidden to crystal patterns. Now it was simple for anyone to see that the number of possible forms of matter had gone from being strictly limited to unlimited. This was a major paradigm shift.

The "projection method" was another important idea developed by several independent groups of theorists. According to this approach, the Penrose tiling and other quasiperiodic patterns are obtained by projections or "shadows" of a higher-dimensional periodic packing of "hypercubes," which are the equivalent of three-dimensional cubes, but in imagined geometries with four or more dimensions of space. Most people cannot visualize how the method works without advanced training, but mathematicians and physicists find the concept very powerful for analyzing the atomic structure of quasicrystals and computing their diffraction properties.

The generalized multigrid and projection methods are powerful mathematical tools for generating patterns of rhombus tiles in two dimensions or rhombohedrons in three dimensions. But they have a major limitation: They provide no information about matching rules. For example, the patterns with eleven-fold symmetry (see the color insert, image 1) and with seventeen-fold symmetry (as shown below) were generated by the multigrid method.

These beautifully intricate patterns are composed of simple rhombus shapes: some fat, some medium, and some thin. But there are no notches or interlocks to prevent the tile shapes from being arranged in a crystal pattern.

So if you were given a pile containing these tile shapes and asked to cover a floor with them without using pictures of the complete pattern as a guide, you might wind up with an ordinary crystal pattern because they are so simple to construct. You might also make a random pattern. But the chance that you would be able to make a quasicrystal pattern is very small. You would need matching rules to guide you, and to help in recognizing if you made a mistake along the way.

Imagine replacing each different tile in the pattern opposite with a group of atoms. Even though the precisely ordered quasicrystal arrangement would be possible, it seems intuitively much more likely that a liquid would tend to solidify into either a crystal or a random arrangement if there were no interactions between atoms that could act as matching rules to prevent that from happening. There are many more of those arrangements and each of them requires less delicate coordination of the atoms than a quasicrystal.

That was why Dov and I had originally worked so hard to show that it is possible to construct interlocks for our fat and skinny rhombohedrons which act as matching rules that prevent both crystal and random arrangements and force quasicrystal ones.

But were matching rules enough to explain why quasicrystals form? I wondered. Maybe a quasicrystal would need additional properties in order for the atoms to organize naturally into the ideal quasiperiodic arrangement.

PRINCETON, JANUARY, 1985: Josh Socolar volunteered to work with me on this challenging question. He had already proven his talents in our previous work generalizing the multigrid approach to arbitrary symmetries, so I was delighted that he wanted to take on a bigger project. Josh was tall and lanky, and always managed to convey a sense of patience and thoughtfulness, which was unusual for someone so young. I felt like I was always the one apt to get overly excited and that Josh brought a sense of calm to the discussion. He also had a remarkable geometrical intuition that would prove to be invaluable in all of our collaborations, which have been fruitful and continue to the present day.

Josh and I decided to go back to the Penrose tiling for guidance.

We noted that Penrose's matching rules for two-dimensional patterns included two other properties that the fat and skinny three-dimensional rhombohedrons Dov and I studied did not share. The first missing element was Ammann lines, the wide and narrow channels that appear when each rhombus is decorated with bars and the tiles are put together in a Penrose pattern. Josh and I decided to incorporate the three-dimensional analog of Ammann lines, which we called "Ammann planes," in our geometrical constructions. The second missing property was deflation-inflation rules, protocols for subdividing the two rhombuses in a Penrose tiling into smaller pieces.

Josh and I conjectured that an alternative set of building blocks with all three properties of matching rules (interlocks), Ammann planes, and deflation-inflation rules might be the secret to understanding how real atoms come together in a liquid to form a quasicrystal. The Ammann planes and deflation-inflation rules might be important in explaining how atoms beginning from some random arrangement organize into a precise quasiperiodic arrangement, and the interlock rules Dov and I developed might be important in explaining how they remain locked in that configuration.

The subtle reasoning was as follows: If the building blocks could be construed as lying along quasiperiodically spaced Ammann planes, then it would be possible to imagine a liquid solidifying into a quasicrystal beginning with some small seed cluster of atoms to which more atoms would attach one layer at a time. Each layer would correspond to an Ammann plane.

The layer-by-layer growth would be analogous to the way many periodic crystals form, so it was reasonable to imagine that something similar occurs for quasicrystals.

The three-dimensional deflation-inflation rules seemed to suggest another way that quasicrystals could grow. First, atoms in a liquid might form many small clusters; then those clusters could come together to make larger clusters; then larger clusters could come together to form yet larger clusters; and so on. This hierarchical clustering of smaller bits to make larger bits might correspond to the way small tiles combine together into larger tiles according to the deflation-inflation rules.

We also imagined that some quasicrystals might solidify by using a combination of layer-by-layer and hierarchical growth.

The fat and skinny rhombohedrons Dov and I had constructed with our cardboard cutouts had interlock rules, but nothing like Ammann planes or deflation-inflation rules. The challenge for Josh and me was to find another set of building blocks that had all three properties. To accomplish that for the complicated case of icosahedral symmetry in three dimensions would be a significant mathematical feat, comparable to what Penrose had accomplished with his two-dimensional designs. But if we succeeded, we could explain that growing quasicrystals in a liquid can be as simple and natural as growing ordinary crystals.

But did building blocks exist that possessed all three properties?

Josh and I set out to find the answer. Shortly after the first papers on quasicrystals appeared at the end of 1984, we began to work intensively on a new mathematical approach to generate quasicrystals based on lessons learned from Penrose tilings.

Our approach involved a weird combination of algebra using pencil and paper and physical geometrical constructions in three dimensions. Algebraic equations had to be solved to predict the precise positions of the Ammann planes in three dimensions, which was my job. Josh would then see where the Ammann planes intersected and use our generalized multigrid method to determine the shapes of the building blocks and how the Ammann planes would pass through them.

The fact that we were working in two different physical locations made the project even more challenging. Josh was at the University of Pennsylvania in Philadelphia, but I was still continuing my sabbatical leave and had become a visiting fellow at the Institute for Advanced Study in Princeton, New Jersey. Skype would not be invented for nearly two more decades. So Josh and I could only communicate over the phone and could not exchange any images.

I would telephone Josh and describe how my algebraic calculations dictated how the Ammann planes should be arranged. He would then describe to me the building blocks implied by my calculations. Josh was able to combine our separate ideas and construct some truly remarkable physical models out of transparent, colored plastic sheets that remain fixtures on my office shelf today. When I finally saw the models several weeks later, I was excited to see that our two calculations fit together hand in glove. We submitted our paper to *Physical Review B* in September of 1985. There was no question that we had solved the problem.

Now we knew for sure that there were building blocks for three-dimensional icosahedral symmetry with interlocking matching rules, Ammann planes, and deflation-inflation rules. They had all the same properties of two-dimensional Penrose tiles but with much more complex symmetry. The work was directly relevant to explaining real-life quasicrystals with icosahedral symmetry.

Josh and I ultimately found a manufacturing company that could fabricate the four types of building blocks we had invented to solve the problem. The plastic blocks have specially designed LEGO-like connections, which serve as substitutes to enforce all of our matching rules.

One shape was the same fat rhombohedron that Dov and I had used, represented by the white blocks seen in the color insert (image 2). The other three shapes were different from any of the ones Dov and I originally studied. They have complicated Greek names based on the number of facets, all of which are rhombuses of the exact same size and shape. The actual names are not so important, but, for those who enjoy practicing their Greek, they are, in increasing order of size: rhombic dodecahedron (twelve rhombus facets, yellow), rhombic icosahedron (twenty rhombus facets, blue), and rhombic triacontahedron (thirty rhombus facets, red).

I have to admit that I enjoy the fabricated units. They not only illustrate how the new building blocks fit together, they also represent a big improvement over the arts-and-crafts experiments Dov and I had first performed with Styrofoam balls and pipe cleaners, and then with cardboard cutouts and magnets.

A few layers showing how the four three-dimensional shapes fit together are shown in the color insert (image 2).

This mathematical tour de force made me feel much more secure that there were no theoretical roadblocks that would prevent us from extending the concept of quasicrystals from the abstract world of two-dimensional Penrose tilings to the realistic world of three-dimensional matter.

Our construction was timely because, by the spring of 1985, the discovery of quasicrystals had launched a hot new field of research. News about new experiments, new potential quasicrystal alloys, and new theoretical ideas from diverse groups all around the world seemed to come out every week. The excitement led to a continuous stream of conferences, workshops, and invited lectures, including the lecture at Caltech that led to my deeply satisfying encounter with Richard Feynman.

It was during this same period of time that Dan Shechtman invited me to visit his laboratory at the Technion in Haifa, Israel. We had met previously at a conference, but only had time for a brief exchange. My visit to Haifa was our first opportunity to spend a substantial amount of time together exchanging ideas.

Dan was a gracious host. He was proud of his work and his nation. He showed me his lab and latest data, and then took me on a tour of the Haifa region all the way up to the Golan Heights.

I admired the courage and independence of mind that had led Dan to make his great discovery. I was not satisfied with our scientific discussion, though. Dan's expertise was in electron microscopy and diffraction, and he had limited interest in theory. It soon became clear to me that he was still enamored with the idea that Ilan Blech had originally proposed to explain his Al_6Mn alloy, which was that the material consisted of icosahedral clusters whose orientations are, for some inexplicable reason, all aligned in the same way despite the fact that their positions in space are random. For some reason, Dan seemed to consider Blech's idea to be equivalent to our quasicrystal theory.

I tried to explain to him the key differences: The Blech model was incomplete since it had large gaps between clusters that were not accounted for; it was not a stable configuration; it would therefore not represent a new phase of matter; nor did it have a diffraction pattern consisting of sharp pinpoints aligned along straight lines.

But I could tell that Dan was not impressed by these distinctions. He apparently thought the Shechtman-Blech model composed of randomly placed icosahedral clusters was easier to imagine and did not seem to consider the important differences that I was pointing out. I felt badly that I had failed to convince him to change his opinion. In fact, he would continue to use the Shechtman-Blech picture rather than the quasicrystal model in his presentations for many years to come.

Shechtman was not the only one resisting the quasicrystal picture.

Within a few months, other plausible alternative explanations for the strange aluminum-manganese alloys would begin to surface. Even more disturbingly, a serious problem with the quasicrystal concept was about to be exposed.

Alternate theories. Conceptual problems. Much to my dismay everything would soon lead to a growing consensus within the scientific community that quasicrystals were, as I had been told all along, impossible.

SIX

PERFECTLY IMPOSSIBLE

PHILADELPHIA, 1987: More than two years had passed since Dov and I had published our paper introducing the concept of quasicrystals. During this period, attitudes toward the concept had undergone a series of mood swings in the scientific community.

For the first year after our paper appeared, the quasicrystal theory was embraced as the only viable scientific explanation for the newly announced alloy with icosahedral symmetry. In fact, the idea seemed to catch the scientific world by storm and triggered a marvelous series of new discoveries.

Scientists began combining aluminum with elements other than manganese, the components of the original experiment, and discovered even more quasicrystal alloys with the symmetry of an icosahedron. In the process, they found a material with eight-fold symmetry, another with ten-fold, and another with twelve-fold, firmly establishing the existence of matter with yet other symmetries previously thought to be impossible.

I admired what all the other scientists were accomplishing. And so far, everything was in agreement with what one would expect according to the quasicrystal theory we had presented. But the good news would not last much longer. The pendulum began to swing in the other direction as competing explanations surfaced, along with serious criticisms.

The first and most vociferous critic was two-time Nobel Laureate Linus Pauling. Pauling was a towering figure in the scientific community. As one of the founders of quantum chemistry and molecular biology, he was widely regarded as one of the most important chemists of the twentieth century.

"There is no such thing as quasicrystals," Pauling liked to joke derisively. "Only quasi-scientists."

Pauling proposed that all the peculiar alloys that had been discovered were complex examples of multiple-twinned crystals, similar to what senior scientists at the National Bureau of Standards had originally suggested. But Pauling had a very different and very explicit atomic arrangement in mind that he claimed could explain the diffraction pattern.

If Pauling was right, there would be nothing newsworthy about any of the new materials. All of our work would sink into obscurity as nothing but a historical curiosity. For those in the materials science and chemistry fields, like Dan Shechtman and his colleagues, Pauling's objections were frightening and considered a serious threat. During the course of his scientific career, Pauling had consistently challenged and prevailed over conventional wisdom. He was not someone you wanted as an intellectual opponent.

I did not share their concerns, though, for one simple reason. I never thought Pauling's alternative proposal was plausible. For one thing, Pauling's multiple-twinning model of the icosahedral Al_6Mn alloy was much more complicated than our quasicrystal explanation. And when it comes to science, the simplest explanation is usually the best.

Josh Socolar and I had established that the quasicrystal picture required four different building blocks (as shown in the color insert, image 2), each composed of tens of atoms, arranged in a quasiperiodic sequence. Pauling believed the materials were, instead, an intergrowth of many crystals oriented at different angles, a version of the multiple-twin idea that John Cahn had originally discussed with Dan Shechtman. According to Pauling's theory, the repeating building blocks in each crystal had over eight hundred atoms per building block. To say it was more complicated than our theory would be a massive understatement.

I was actually becoming much more concerned about another competing idea that had begun to gain prominence around the same time Pauling was making his ideas known—the icosahedral glass model, a theory developed by Peter Stephens from Stony Brook University and Alan Goldman from Brookhaven National Laboratory. This was a much improved version of the Shechtman-Blech model.

The new icosahedral glass model proposed an atomic structure composed of icosahedron-shaped clusters arranged in a disordered pattern in space. That feature explained the word "glass" in the theory, because "glass" refers to materials with random atomic arrangements. In this model, the corners of each icosahedron-shaped cluster were aligned so that they pointed in the same directions in space. That feature was similar to the Shechtman-Blech model, with a notable improvement. Stephens and Goldman included an explanation of how to join the clusters together in a way that resulted in much smaller gaps and cracks.

The two models, icosahedral glass and our quasicrystal theory, could be distinguished, in principle, based on the sharpness and alignment of spots predicted for the diffraction pattern. A perfect quasicrystal produces a pattern of true pinpoints arranged in crisscrossing straight lines. The diffraction pattern for an icosahedral glass was predicted to be very similar, except the spots were fuzzy and did not align perfectly.

Unfortunately, the initial data was ambiguous because of the nature of the material being tested. To put it simply, Shechtman's original aluminum-manganese sample was not very good quality. There was something inherently flawed about the alloy. Groups that had been trying to independently duplicate the sample were encountering the same problem.

The problem with the fuzziness and positions of the diffraction spots observed in the original sample had not been immediately apparent from the published photographs. Such images tend to be overexposed in ways that hide the flaws. But the issue was made glaringly obvious in the more precise X-ray diffraction patterns later made by Paul Heiney and Peter Bancel at Penn's Laboratory for Research on the Structure of Matter.

Their X-ray diffraction lab was just across the street from my office at Penn, so I was able to study the test results as they were being completed. As someone who strongly believed in his own theory, I have to admit that I found the new diffraction images somewhat alarming. They clearly showed that the X-ray diffraction spots were fuzzy and imperfectly aligned, which did not match our predicted pattern. The results appeared to be a match to the competing glass model.

Things looked bad. But even so, I knew that the X-ray results would not necessarily sound the death knell for our theory. There might be a simple explanation for the fuzziness and small misalignments of the diffraction peaks. That was something that would naturally occur if the initial liquid mixture of elements being used to create a quasicrystal was cooled too quickly. The rapid cooling process tended to freeze in randomly placed defects and prevent the atoms from reaching the ideal arrangement.

As it turned out, all of the icosahedral Al_6Mn samples that existed so far had been synthesized using a rapid cooling process. And with good reason. Whenever the material was cooled more slowly, it failed to form a quasicrystal. Instead, the aluminum and manganese atoms would totally rearrange themselves into a classic crystalline arrangement.

Joshua Socolar and I teamed up with the renowned condensed matter theorist Tom Lubensky to analyze the situation. The three of us developed a detailed theory to describe various distortions expected to occur in quasicrystal diffraction patterns because of defects introduced by the rapid cooling process. We found that the predictions could produce the same exact fuzziness and displacement of diffraction peaks that had been observed in X-ray experiments on Al_6Mn. That meant our theory could be adapted to predict either sharp or fuzzy pinpoints, depending on the cooling process. So we were still in the running.

The icosahedral glass model was also still in the running, though, because it predicted fuzzy spots. To make matters worse, the data also allowed room for a version of Pauling's idea of a multiple-twinned crystal, provided one allowed for at least eight hundred atoms in each repeating building block.

So in essence, all three models could explain Shechtman's data.

In principle, there was yet another type of test that could settle the competition, one that involved heating instead of cooling. If one were to heat the sample gently over a long period of time, but not to such high temperatures as to make it melt, three different outcomes were possible. It would either form a more perfect quasicrystal with sharp peaks, as Dov and I had predicted, or it would form a more perfect multiple-twinned crystal in accordance with Pauling's theory, or it would remain a disordered icosahedral glass with fuzzy peaks, in accordance with the Stephens-Goldman model.

But unfortunately, the heating test could never be performed with Shechtman's alloy of Al_6Mn because of its tendency to crystallize. Heating the alloy for even a short amount of time would destroy the icosahedral symmetry altogether, making it impossible to determine which theory was right.

In fact, more than three decades since its discovery, experiments still cannot definitively determine whether Shechtman's Al_6Mn material is a genuine quasicrystal, an icosahedral glass, or one of Linus Pauling's multiple-twinned crystals.

That dilemma partially explains why it took so long for the scientific community to accept quasicrystals as a new form of matter.

The second reason for the delay in acceptance was more theoretical in nature, based on a thorough study of Penrose tiling. Critics who preferred the icosahedral glass picture argued that a true quasicrystal was an unattainable state because there was no plausible way to "grow" one.

To crystallographers, the word "grow" means forming crystals slowly from a liquid mixture of atoms. One can make sugar crystals, commonly known as rock candy, by dissolving lots of sugar in water and then waiting a few days for the sugar crystals to form. Similar processes occur in nature and in the laboratory. What occurs on a microscopic scale is that, beginning from some small cluster of atoms in the liquid, more and more atoms attach until the cluster "grows" to a visible size. For this to occur, it is important that the atoms maintain a regular periodic order whenever they attach. Since atoms in the liquid randomly approaching a cluster only interact with the nearest atoms in that cluster, there must be simple forces or, equivalently, simple rules that determine where atoms attach and where they do not.

Common experience in constructing Penrose tilings suggested that simple "growth rules" like this do not exist for quasicrystals. Imagine you decided to cover a large surface with a Penrose pattern using a pile of fat and skinny rhombus tiles. You know about matching rules, so you would make sure that any tile you added would join in accordance with the matching interlocks Penrose prescribed. Your goal is to completely cover the surface without leaving any gaps.

You might guess that this could be easily accomplished. After all,

Penrose showed that it is possible to completely cover a surface, even one of infinite extent, using his interlocking tiles.

But you would be dead wrong. The Penrose tiling is like a challenging jigsaw puzzle composed of only two shapes. There is a valid solution to the problem, a way in which all of the jigsaw puzzle pieces can interconnect. But finding that precise solution requires patience and a lot of trial and error.

If you started putting the tiles together one by one, chances are that you would run into difficulty after only a dozen or so pieces even if you meticulously followed all of the interlock rules each time you added a tile. You would eventually run into a spot where neither a fat nor a skinny tile could fit. You could start over and try again making different choices. Odds are, though, that you would not get much further.

The problem is that Penrose interlock rules only ensure that an added tile is properly aligned with its immediate neighbors. The rules do not ensure that the added tile is properly aligned with respect to the rest of the tiles in the pattern. So unless you are lucky, some of the tiles already added to distant parts of the pattern will be in conflict. And that conflict only becomes apparent when you suddenly reach a point where no tiles can fit. Scientists call that type of dead end a defect.

If you continued adding tiles, you would soon find yourself creating another defect. And then another and another and another. By the time you put together hundreds of tiles, you would have so many defects that you would hardly recognize the result as a Penrose tiling.

Of course, Penrose proved that it was possible for the tiles to be arranged in a perfect, gapless pattern. But he never claimed that one could construct a pattern by putting his tiles together in an arbitrary order. In fact, he was well aware of the fact that the proper arrangement was nearly impossible to find.

If that problem occurred for Penrose tiles with matching rules, critics argued, the same must occur for atoms attaching one by one to a cluster to form a quasicrystal: So many defects would form during growth that it would be nearly impossible to form anything resembling a true quasicrystal. Skeptics concluded that for all practical purposes, a perfect quasicrystal was an unattainable state of matter.

This was a true low point in the quasicrystal story. The two problems were seemingly insurmountable. The best experiments on Al_6Mn were being conducted with a rapid cooling process that always created fuzzy spots in the X-ray diffraction pattern, instead of the sharp pinpoints we had predicted. And now, there appeared to be a strong conceptual argument that quasicrystals were effectively an impossible state of matter.

The debate was resolved by two breakthroughs. One theoretical and the other experimental.

YORKTOWN HEIGHTS, JULY 1987: The theoretical breakthrough came with the discovery of an alternative to Penrose's interlocking rules, which we called "growth rules." They made it possible to add tiles one by one to a pattern without making any mistakes or creating any defects. The growth rules were inspired by yet another visit to the IBM Thomas J. Watson Research Center in Yorktown Heights, New York. This time I had been invited to spend the summer continuing my research on quasicrystals.

One day while working at the Center, a researcher named George Onoda invited me and his colleague David DiVincenzo to lunch. He wanted to discuss a new idea about how to avoid defects in Penrose tilings. I had known George for several years. We met during my first sabbatical at IBM in 1984, around the same time that Dov and I published our first papers on quasicrystals. I had known David since he had been a graduate student at Penn.

When we sat down to lunch, George explained that he was familiar with the problem of frequent defects created by following the Penrose interlocking rules. He had struggled with the issue and found that he could construct additional rules that could ensure defects were made less frequently. That idea sounded intriguing. So we quickly finished our lunch and moved to a nearby meeting area where we could sit around a big circular coffee table. George took out a box full of paper Penrose tiles and began demonstrating his new rules.

Sure enough, George's rules were an improvement. We still encountered a dead end, a gap that could not be filled, but we were able to put together two dozen tiles before that first occurrence. Once we understood how George's new rules worked, we noticed that we could add yet another rule that would make the process even better. And after we tried that rule, we discovered yet another rule that would lead to even more improvement. Each of us took turns adding new rules over the next two hours until, suddenly, we found that we could cover the entire table with George's tiles without making any mistakes or adding any more rules.

It must have been a strange sight to see three scientists hunched over a table intent on constructing a homemade paper puzzle. But if anyone happened to take exception, we would have never noticed. The more time we spent on the project, the more absorbing it became. None of us had ever expected to find rules that made it possible to put together so many Penrose tiles without making a defect.

Unfortunately, we only managed to achieve this success after drawing up a long laundry list of seemingly arcane rules, things like "if a configuration like such-and-such occurs, add a fat tile to this particular edge of it." But then, as I began to study the list more closely, I noticed that the entire list of rules could be restated compactly if expressed in terms of adding tiles to something called an "open vertex."

A vertex of a tiling is any point where the corners of several tiles meet. An open vertex is where there remains a wedge of space to add more tiles.

The long list of new rules we devised were able to be reduced to one sentence: Only add a tile to a vertex if there is a unique choice that produces a legal vertex, one that is allowed in a perfect Penrose tiling; otherwise, randomly choose another vertex and try again.

Could such a simple rule really work? Proving it mathematically was a challenge that took several months. Once again, I contacted Josh Socolar, who was by now considered a world expert on tilings. It had been a few years since the two of us had first theorized why quasicrystals might form, using matching interlock rules, Ammann lines, and deflation-inflation rules. Now, using a brilliant combination of computer programming and mathematical reasoning that Josh had devised, we were able to demonstrate that all three properties were essential to proving that the new vertex rule worked without fail, with the minor technical twist that the initial seed cluster of tiles included a configuration known to Penrose tilers as a "decapod."

Our new growth rules were fundamentally different from Penrose's original matching rules. Those rules constrain the way *two tiles* can join along an *edge*. The growth rules constrain the way a *group of tiles* can join around a *vertex*. Just like the matching rules, though, the growth rules could result from realistic atomic interactions in which the forces between atoms only extend across a few atomic bond lengths.

The growth rules surprised the scientific community. Among those most astounded was Roger Penrose. I first met Roger in 1985 when I invited him to Penn to meet with both my theory group and my experimental colleagues working on quasicrystals. I eagerly showed him all the research that his ingenious invention had inspired. Roger was the epitome of modesty and graciousness. With his crisp British accent, he politely asked hundreds of questions and generously shared his own ideas. We quickly established a great rapport which continues to this day, as we share interests in both quasicrystals and cosmology.

In 1987, though, Roger was still convinced that the skeptics were right. Based on the problems encountered while constructing Penrose tilings, he believed it was impossible for atoms with ordinary interatomic forces to form highly perfect quasicrystals. A few years later, though, he changed his mind. In 1996, I was invited to a 65th birthday fest at Oxford University to honor Roger and his many historic contributions. The event gave me the opportunity to show Roger the mathematical proof of our growth rules. As a memento, I gave him a rare set of our three-dimensional building blocks (color insert, 2), which he gratefully accepted.

It would take us nearly three more decades before we could complete the proof for growth rules in three dimensions. Although the same principles apply as for two-dimensional Penrose tiles, arriving at a proof is much more difficult. It is much harder to visualize three-dimensional building blocks and there are many more configurations to consider. Josh and I put the problem aside until 2016, when we decided to revisit it using improved visualization techniques. By then, Josh was a professor at Duke University and was joined by his talented undergraduate student, Connor Hann. Together, the three of us finally completed the proof.

Finding growth rules for Penrose tilings in two dimensions had been enough to decimate the skeptics' conceptual argument that a perfect quasicrystal was an unattainable state. *But would it ever be possible to find a combination of elements that would form a perfect quasicrystal in the laboratory?*

SENDAI, JAPAN, 1987: Even before our paper on growth rules was published, a scientist a quarter of the way around the world from us solved that problem.

An-Pang Tsai and his collaborators at Japan's Tohoku University announced the discovery of a beautiful new icosahedral quasicrystal composed of aluminum, copper, and iron. Unlike quasicrystals synthesized earlier, Tsai's sample did not require rapid cooling. As a result, it could be annealed, meaning it could be heated gently for days without transforming into a crystal. The annealed quasicrystal was nearly defect-free and had a solid, beautifully faceted shape that clearly showed its inherent five-fold symmetries.

The image, which appears below, may appear ordinary at first glance, like a faceted diamond or quartz crystal. But this is far from ordinary. These were the first-ever absolutely perfect pentagonal facets ever seen, and a major scientific advance compared to the disordered, feathery structures formed by Shechtman's Al_6Mn alloy.

Before the discovery of quasicrystals, most scientists would have declared facets with five-fold symmetry were impossible because they violated the centuries-old rules established by Haüy and Bravais. Yet here is indisputable evidence that they exist.

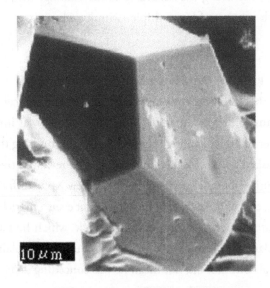

10 μm

It took some time to verify the results. But Paul Heiney and his student Peter Bancel were finally able to take an X-ray diffraction pattern for the new aluminum, copper, and iron sample, just as they had done for Shechtman's Al_6Mn. This time, Heiney and Bancel found something impressively different. The Bragg peaks in Tsai's sample were sharp and pinpoint, not fuzzy, and the positions of those peaks perfectly aligned along straight lines, in agreement with our predictions for the icosahedral quasicrystal model.

At last, here was the first unambiguous, bona fide case of an icosahedral quasicrystal. Proponents of the icosahedral glass model graciously conceded, and quasicrystals were finally accepted as a true form of matter. Over the next few years, many more examples of perfect quasicrystals were found, in many cases by Tsai and his collaborators. When I finally had the chance to meet him in Japan many years later, I was pleased to be able to express in person how much I appreciated and admired his historic contributions.

Despite the new experimental proof, there remained a smattering of skeptics, including the venerable Linus Pauling, who maintained his staunch support of his multiple-twinning idea.

PHILADELPHIA, 1989: I invited Pauling to visit me at Penn in order to review the definitive measurements Heiney and Bancel had made of Tsai's sample. It was a memorable occasion, and I was impressed by the number of hours Pauling spent meticulously combing through the data. He asked a lot of detailed questions as he reviewed the data, trying to identify potential problems with the new X-ray diffraction test.

By the end of the day, Pauling agreed that even a model with eight hundred atoms per repeating building block, which he had claimed could explain Shechtman's Al_6Mn data, would not explain the new quasicrystal. But that did not mean he was conceding defeat. It meant he would go back and substantially increase the number of atoms per building block in his theory until he could fit the new data, even though that would make his theory even more byzantine than before.

Pauling told us that he planned to write a new article in the *Proceedings of the National Academy of Sciences* in which he would describe his revised multiple-twinning model for Tsai's perfect quasicrystal. In a gesture of professional respect, he invited us to write a companion piece explaining why the quasicrystal model explained the result more simply. With Pauling's support, both articles appeared back to back in an issue later that year.

Pauling and I continued to correspond over the years as more and more combinations of elements producing perfect quasicrystals were found in the laboratory. As the years passed, he became increasingly familiar with the quasicrystal picture and seemed to acknowledge its advantages. I believe he understood that the quasicrystal theory would prevail, but he was not ready to give up his favorite idea. I could appreciate that. I enjoyed our ongoing friendly debate and was saddened to read the news in 1994 that he had passed away at the age of ninety-three.

By this time it was clear that there were no remaining roadblocks to synthesizing perfect, stable quasicrystals in the laboratory. The subject had achieved such wide acceptance that there were now annual international meetings on quasicrystals, with hundreds of people and creative contributions from a wide spectrum of experimentalists, theoreticians, and pure mathematicians.

I was proud to have been part of it. But I also felt that the subject was becoming too crowded and too mature for my tastes. In order for me to continue to work on quasicrystals, I needed to pursue a question that no one else was considering.

I reminded myself that growing perfect quasicrystals in the laboratory had proven to be easier than anyone had ever thought.

Might it be possible for perfect quasicrystals to develop on their own without any human intervention?

That idea was reminiscent of the question I had briefly explored in 1984, shortly after Dov and I had published our first paper: If synthetic quasicrystals were possible and so easy to create, what about natural quasicrystals?

Thus far, quasicrystals had only been synthesized in the laboratory under carefully controlled conditions that were too pristine to ever be duplicated in the natural world. So I was pretty sure that other scientists would consider natural quasicrystals to be absolute folly. Impossible. And that was a good enough reason for me to begin to pursue the idea.

THE QUEST BEGINS

SEVEN

DID NATURE BEAT US?

PRINCETON, 1999: "Has anyone ever discovered a *natural* quasi-crystal?"

The jovial white-haired fellow rushed up to the lectern to ask the question the moment my lecture ended. I had just joined the faculty in the Department of Physics at Princeton and had decided to focus my debut talk on the history of quasicrystals. It had now been fifteen years since Dov Levine and I had first introduced the concept.

I did not recognize the man asking the question from any faculty meetings and soon understood why. He introduced himself as Ken Deffeyes from the Geosciences Department. I was surprised that he was attending the talk. Generally the only attendees at the weekly colloquiums were physicists and astrophysicists.

I appreciated Ken's question because it meant that he understood the point of my lecture. I had presented a series of new theoretical arguments showing that quasicrystals can be as stable and as easy to grow as crystals. So it was only logical that, as a geologist, he would want to ask if they were known to exist in the natural world.

"No," I answered. "I have spent time looking haphazardly in museum collections in the past without success. However," I added with

a smile, "I have an idea for a systematic way of searching for them."
Ken's eyes grew wide, and he asked me to describe the idea.

I told him that it involved an automated search through a computer database containing tens of thousands of diffraction patterns.
Some of the patterns were from synthetic materials. But nearly ten
thousand of them were from natural minerals. Several years earlier,
I had hired an undergraduate to search through the database pattern
by pattern for potential quasicrystals. But he had run out of energy
after a short period of time. Later, I realized that the screening process
could probably be entirely automated. One could narrow the search
with a computer program, obtain samples of the most promising candidates, and test them in the laboratory.

Ken thought that was a great idea and told me he knew just the
person for the job, a bright undergraduate named Peter Lu. Peter had
won national gold medals in the "Rocks, Minerals, and Fossils" event
at four consecutive National Science Olympiad tournaments in high
school. He was currently a junior in the Physics Department, Ken explained, which meant he would be looking for a senior thesis project
the next year. Peter also had experience with an electron microscope,
which would be an asset in case there were any potential quasicrystals
identified during the search.

Ken also recommended that I contact Nan Yao, the director of
Princeton's Imaging and Analysis Center and a specialist in electron
microscopy. Ken said that Nan was a gifted teacher who had trained
Peter. Nan was also highly skilled in obtaining diffraction patterns
from unusual materials.

The next day, Ken introduced me to Peter, who seemed to be perfectly suited for the project. Peter was an intense, ambitious student
looking for a challenge. He was short and youthful, but spoke with
enormous self-confidence and in an authoritative tone. He had not

attended my lecture but felt that he had heard enough from Ken to speak assertively about the project and his qualifications.

Peter and Ken then took me to meet Nan at the Imaging Center and tour the facilities. The Center contained electron microscopes and an array of other expensive instruments for studying a variety of materials. It served scientists and students in departments throughout the university as well as experts at nearby industrial laboratories. Nan was enthusiastic about our project and eager to help in any way possible, which included making sure we were allotted time to work with the Center's electron microscope. I took note of his calm reserve and expertise as he showed us around the facilities. I knew he would be a valuable member of the team.

With Ken, Peter, and Nan on board, I found that I had the right combination of people, knowledge, and skills to move forward with a systematic search for natural quasicrystals. So my long-awaited quest began in earnest.

Although Peter's talents were largely in mineralogy and experimental physics, he rapidly absorbed the basics of quasicrystal mathematics. We began working on a computer algorithm that would enable us to rank the likelihood that a given candidate mineral is a quasicrystal based on its diffraction pattern, as recorded by the International Centre for Diffraction Data (ICDD).

The ICDD is a nonprofit organization that collects information about materials and their X-ray powder-diffraction patterns from laboratories all over the world. The information is compiled in an encrypted database, and scientists and engineers purchase subscriptions to gain access to it. Experts commonly use the database to compare diffraction patterns they are examining with patterns from previously known materials.

The ICDD also provides software to extract information from

the database, but their program turned out to be too cumbersome for our purposes. It could only provide access to one powder-diffraction pattern at a time, along with a lengthy amount of descriptive information that was superfluous for our purposes.

In order to conduct our statistical analysis, we only needed access to the powder-diffraction data. So we wrote to the ICDD, explained our project, and asked if they would allow us to work with a decrypted version of their database. We could then write our own software to extract the relevant information and compress it into one large file for our analysis. We were not sure what to expect, because we were asking for special access to their most valuable commodity. But they generously provided us with everything we needed at no cost.

The next hurdle for us to overcome was that we were limited to working with *powder*-diffraction patterns. If the ICDD had been able to offer *single-grain* diffraction patterns, it would have been an afternoon's work to pick out the quasicrystal patterns (below left) from the crystal patterns (below right).

The ICDD does not collect single-grain diffraction patterns because they do not exist for most materials. One needs a sample of a certain size and thickness to make a high-quality single-grain diffraction pattern. For most of the minerals and materials that scientists

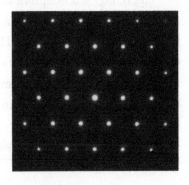

study, those types of samples are too difficult or too time-consuming to find.

Instead, scientists collect many tiny individual grains oriented at random angles with respect to each other. A "powder" of grains like this might occur naturally, or it could be easily prepared by grinding one or more small samples into a fine powder.

Shining X-rays on the collection of grains produces what is called an "X-ray powder diffraction pattern," which combines the diffraction patterns from *all* the grains. For example, if all the individual quasicrystals had a sharp, pinpoint diffraction pattern like the one shown on the left below, the *powder*-diffraction pattern would look like the one on the right.

The powder pattern is similar to what you might observe if the sharp pinpoint diffraction patterns were put on a turntable and spun rapidly so that each point became a circular blur. The left pattern shows points arranged with clear ten-fold symmetry. In the powder pattern on the right, all of the information about symmetry is lost. All that remains are rings with different radii and intensities.

Imagine that you only had the image on the right. Could you reconstruct the fact that it came from a powder of randomly oriented grains in which each individual grain produces a pattern like

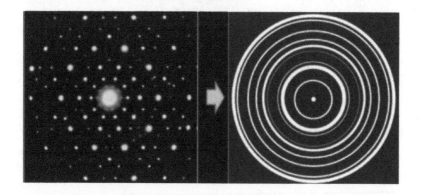

the one on the left? That was the question we were trying to answer. Amazingly enough, Peter and I were able to determine that there was enough information in the spacings and intensities of the rings on the right to identify potential quasicrystals and to infer the now-familiar snowflake pattern on the left.

The plot below summarizes our findings. The graph compares two different properties that we computed for each powder pattern in the ICDD catalog. The horizontal axis measures how close the powder-diffraction rings for the sample are to the ideal radii for a perfect icosahedral quasicrystal. The vertical axis measures how well the intensities match.

The two shaded squares in the lower left-hand side of the graph

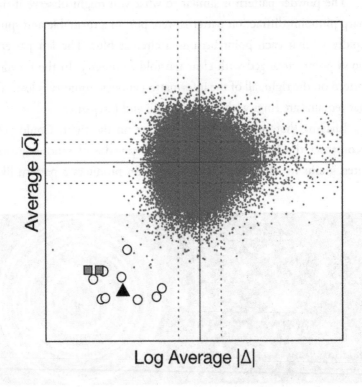

represent two of the known synthetic quasicrystals that were already in the ICDD catalogue. So in practice, those two squares were as close as one could get to perfection. If the powder pattern of a natural mineral had scored close to those squares, we could reasonably expect that it was a quasicrystal in which each grain had a pinpoint diffraction pattern.

The points on the graph represent the results for more than nine thousand minerals that lie too far from the shaded squares to be considered promising candidates. The circles represent the mineral powder patterns that came the closest to the squares and indicated potential quasicrystals.

The circles corresponded to the mineral samples that Peter and I now had to track down and bring to our lab in Princeton for further study. Once a sample arrived, it would be sliced into thin sections and examined under the electron microscope to determine if it was truly a quasicrystal.

When Peter's final year at Princeton was finished, he presented the results of his work in his senior thesis defense. According to tradition, a team of faculty members grill the senior to test the student's familiarity with the subject. But Peter opted for a bit of levity and decided that he was the one who was going to do the grilling. Literally.

So as part of the formal defense of his thesis, Peter amused the attendees by cooking a raw steak on a special frying pan made with a thin coating of quasicrystal metal. The use of synthetic quasicrystals for a nonstick coating was one of the first commercial applications of the new form of matter. The coating was conceived and patented by a French quasicrystal scientist, Jean-Marie Dubois, and his collaborators. A French manufacturer sold the pans under the trade name Cybernox.

The quasicrystal coating was slippery like Teflon, a popular nonstick coating, but much more durable. Peter was able to fry his steak without using any butter, demonstrating that nothing stuck to the quasicrystal surface. He capped off the demonstration by slicing into the steak with a sharp knife while it was still in the pan, something no one would ever attempt to do in a pan with a Teflon coating. Peter was able to show that there was no damage to the pan because of the hardness of the quasicrystal material. But the same could not be said for the steak knife, which left significant streaks of metal shavings on the pan's surface.

Peter also presented the details of our search through the ICDD catalog. He explained the search algorithm we had developed and described the candidates we had managed to study. We had not succeeded in discovering a natural quasicrystal. But the process of trying to collect and test the minerals was a series of adventure stories, with the occasional funny mishap.

For example, after months of hard work, we had finally managed to obtain a sample of one of our topmost mineral candidates. The sample measured a few inches across. In order to study it under the electron microscope, though, we needed to obtain a slice as thin as a human hair.

The slicing procedure required a special facility that did not exist at Princeton. So we arranged to send the sample to a lab at UCLA. We expected to receive a thin slice from the lab, along with the remainder of the sample. If we succeeded in finding a quasicrystal in the slice, the rest of the sample would be extremely valuable for follow-up studies and would ultimately end up as a prized exhibit in a museum.

But when the package came back from UCLA, I opened the box and found that all it contained was a single ultrathin slice. *What happened to the rest of the rare sample we had worked so hard to obtain?*

I frantically called UCLA to find out when they would be sending us the rest of our material. When I finally reached the technician involved, he cheerfully reported, "Oh, we assumed you only needed the one slice, so we threw out the rest of the sample."

I was horrified. As far as we knew, this might turn out to be the only sample of that mineral in the world. If we examined the slice, and discovered it contained the first natural quasicrystal, we would have to live with the fact that 99.99 percent of the rare material had been tossed in the garbage can. The next few hours were nerve-wracking as we waited for Nan to check the wafer-thin slice. When he reported that the sample was a dud, Peter and I left the lab with a strange mixture of disappointment and relief.

Ultimately, all of the minerals that we identified, collected, and tested turned out to be duds, as well. A year after Peter passed his oral defense exam, we published a paper about our experiment in the *Physical Review Letters,* describing our computer search algorithm and our lengthy string of failures.

A weakness in our approach, we concluded, was that the quality of the data collected by the ICDD from diverse laboratories across the world was uneven. As a result, our automated search algorithm yielded a lot of false positives. I would have to accept the fact that there would continue to be many more failures before finding a true natural quasicrystal.

Peter graduated summa cum laude from Princeton, and headed off for Harvard Graduate School to study completely different topics. Although he was no longer involved in my search for natural quasicrystals, he remained intrigued by the beauty of quasicrystal tilings. When Peter was still an undergraduate, the two of us would sometimes discuss the fact that Penrose had managed to construct a quasicrystal tiling without being aware of its hidden quasiperiodic order. So perhaps, we speculated, it might be possible that quasiperiodic tilings had been unintentionally designed by someone prior to Penrose. A likely place to search for such a thing was among Islamic tilings because many Islamic cultures had an advanced knowledge of mathematics and an interest in geometric patterns.

Years later, when Peter had a chance to travel to Bukhara, Uzbekistan, on his summer vacation, he found many examples of periodic patterns that included ten-fold stars as part of the repeating motif. That experience inspired him to search through catalogs of Islamic tilings when he returned home. Many tilings were like the ones he saw in Bukhara—periodic patterns with regularly placed five- and tenfold stars. But a pattern found on the Darb-i Imam shrine in Isfahan, Iran, a monument with an inscription dated 1453, defied simple description (see color insert, image 3).

Peter contacted me soon afterward to help analyze the complex tiling. We transposed the photograph into a precise geometrical pattern composed of three shapes known as "girih tiles," as shown in image 4 in the color insert. From this, we discovered that the pattern was almost perfectly quasiperiodic, except for a small percentage of errors that may have been due to later repairs. Furthermore, we found that one could construct this pattern and an infinite extension of it using a kind of deflation, or subdivision, rule that was much more complex than the one for Penrose tiles.

Unfortunately, we have no record of how the artisans of Darb-i Imam designed the complex pattern. One can only hazard a guess, based on the fragments observed today on the shrine. Although the design suggests some knowledge of the deflation rule Peter and I identified, there was no evidence that the artisans applied any matching rules. At the present time, there is no other Islamic tiling with as many surviving tiles that is known to be as perfectly quasiperiodic.

The Islamic tiling project was a fascinating digression into art and archaeology, but I was not ready to give up on my quest for a natural quasicrystal. I was still hoping that someone would respond to the paper Peter and I had published describing our search through the ICDD catalog.

In our concluding paragraph, we had offered to share the remaining list of potential candidates that we had been unable to examine with anyone willing to join the search. "Those interested are encouraged to contact PJL and PJS [Peter and Paul]."

We had hoped the invitation would act as a scientific homing beacon. But no one answered our call for help . . . no one . . . not for six long years. And then . . .

EIGHT

LUCA

PRINCETON, BOSTON, AND FLORENCE, ITALY, 2007: On May 31, 2007, Peter Lu and I received an email from an Italian mineralogist named Luca Bindi. We were caught off guard. Neither of us had ever heard of Luca before. But he had obviously heard of us.

Luca had been studying a special class of minerals known as "incommensurate crystals" whose atoms are spaced quasiperiodically, similar to a quasicrystal, but in a way that maintains the long-standing rotational symmetry rules of Haüy and Bravais, which quasicrystals do not.

While researching the topic, Luca had unearthed our paper describing our methodical search for natural quasicrystals. He had taken note of the fact that we were inviting potential collaborators to contact us and was writing to take us up on the offer.

Luca identified himself as the head of the Department of Mineralogy at the Museo di Storia Naturale dell'Università di Firenze, the Natural History Museum at the University of Florence. He volunteered to study any potential quasicrystal mineral candidates that could be found in his museum's collection.

In other words, I thought, here is an Italian scientist I have never heard of, who is volunteering to join a wild-goose chase designed by

a couple of American scientists whom he has never met, and whose search for natural quasicrystals had, by the way, failed to produce any tangible results for the past eight years. *Who is this guy?* I wondered.

Peter was, by then, an advanced graduate student at Harvard working on projects unrelated to quasicrystals. He asked if I thought we should proceed to work with this unknown scientist. *Why not?* I figured.

Almost immediately, Luca became as obsessed about the search for natural quasicrystals as me. Although he was athletic and loved being outdoors, he also had the patience required to spend countless hours working alone in the laboratory, even if the chance of success was minuscule.

Peter and I began by sending Luca our list of prime quasicrystal suspects based on the ICDD catalog. Luca then proceeded to collect samples in his museum that were on our list and carefully analyze them.

It did not go well. Over the next several months, he reported disappointing results to me on a regular basis. Failure after failure after failure.

At one point, I suggested to Luca that, rather than continue to search terrestrial minerals, "Meteorites are more promising because they include a variety of pure metal alloys. I would be interested to work on them with you." That thought would later prove to be prophetic. But Luca did not pick up on my suggestion at the time, perhaps because he was a mineralogist, and meteorites were outside his area of expertise.

It was during this period of back-and-forth communication that Luca and I began to develop a deep friendship, which was tested from the very beginning by the ongoing frustrations he was experiencing in the lab. Our respect for each other was grounded in science and nurtured by daily emails and Skype chats.

FLORENCE AND GENOA, ITALY, NOVEMBER 3, 2008: After more than a year of disappointing results, Luca suddenly, and unilaterally, did what any good scientist would do. He cast aside the losing strategy and adopted a new one.

Although the ICDD file that Peter and I had used for our initial analysis contained diffraction patterns for thousands of minerals, there were some rare and recently discovered natural minerals that were not yet included. Luca decided to specifically target those minerals. He narrowed the search even more by focusing on the ones that included metallic aluminum and copper, which was a popular combination of elements being used at the time to create many synthetic quasicrystals.

In early November of 2008, I went to Italy to attend the annual Festival della Scienza in Genoa. I had been invited to the festival to speak about *Endless Universe*, a book written for the general public about the cyclic theory of the universe that I developed with physicist Neil Turok. The cyclic theory has advanced and improved significantly since. Today it is the leading alternative to the inflationary model, a model that I also helped to develop a few decades earlier but no longer consider viable.

I had not heard from Luca for some time. So I had not bothered to tell him that I would be in Italy that week. As I was walking through the Piazza San Lorenzo in front of my hotel in search of some good Italian coffee, I felt a vibration from my BlackBerry. It was Luca. I opened the message, expecting news of yet another failure. Instead, there was an email which began:

> I studied a museum sample (belonging to the mineralogical collection of Museum) labelled khatyrkite ($CuAl_2$). By means of a preliminary SEM [scanning electron microscope] study, I realized that such a sample actually consisted of four different phases, i.e., cupalite (CuAl), khatyrkite ($CuAl_2$), an unknown phase with composition CuFeAl, and finally a phase with stoichiometry $Al_{65}Cu_{20}Fe_{15}$ (normalized to 100 atoms).

The rest of the message focused on the last phase, the mineral Luca described with the chemical formula $Al_{65}Cu_{20}Fe_{15}$, meaning 65 percent aluminum, 20 percent copper, and 15 percent iron.

The sample he was referring to is shown in the color insert (image 5) in its original plastic box next to a 5-cent euro coin included for scale. Most of what you see inside the box is a lump of putty holding the rock in place so it would not knock around and shatter when the box was moved. The entire sample, a mere three millimeters in diameter (shown enlarged in image 6 of the color insert), was being held in place at its uppermost tip by the wad of putty.

The image with the 5-cent euro coin was my first peek at the tiny grain that was about to launch a grand adventure.

The box described the sample as "Khatyrkite" (pronounced KAT-er-kite), a crystal mineral composed of $CuAl_2$ (one atom of copper for every two atoms of aluminum). Khatyrkite was listed in the International Mineralogical Association (IMA) official catalog, which meant that its composition and periodic structure were already known, and that its properties had been carefully measured and documented. The sample was registered in the official Florence Museum catalog as number 46407/G. The label on top of the plastic box included, for some unknown reason, the number 4061 along with the word "Khatyrka," the name of a river in Far Eastern Russia, and "Koriak Russia," an alternate spelling referring to the Koryak mountain range in Chukotka, the Russian district just north of Kamchatka and east of Siberia.

The close-up (color insert, image 6) revealed that the grain contained a complex conglomeration of minerals. Luca found that the parts with lighter colors included common minerals like olivines, pyroxenes, and spinels. The darker material consisted primarily of alloys of copper and aluminum. The box was labeled "Khatyrkite" because whoever wrote the label considered the crystal $CuAl_2$ to be the component that made the sample worthy of interest.

Luca had already sliced the mineral open to study its composition. He made six delicate sections, each the thickness of a human hair. In order to create the slices, though, Luca was forced to sacrifice the bulk of the sample. Ninety percent of what would eventually turn out to be an extremely precious mineral sample was destroyed in the process. Shown below the close-up of the sample in the color insert is image 7, the thin section that Luca excitedly recounted in his email.

The grayscale image shows a mix of different materials that look like they have been randomly kneaded together. Using an electron microprobe, which bombards a sample with a narrow beam of electrons to measure the chemical composition, Luca was able to identify most of the minerals in the section. Each dot in the image corresponds to a different measurement.

The yellow dots correspond to places where Luca found khatyrkite, $CuAl_2$ (as mentioned, one atom of copper for every two atoms of aluminum). The red dots correspond to another rare crystal known as "cupalite," $CuAl$, a 50-50 mixture of copper and aluminum atoms.

Then, he found some truly puzzling spots. The green dots denote areas with roughly an equal mix of aluminum, copper, and iron atoms, which is a combination that did not appear anywhere in the official IMA catalog of natural minerals. The blue dots are $Al_{65}Cu_{20}Fe_{15}$, another composition that did not exist in the catalog.

Luca was eager to isolate the two mysterious minerals corresponding to the green and blue dots in order to obtain their powder-diffraction patterns and identify them. For this, he took a huge gamble and used a special tool to punch out the regions. The operation took remarkable hand-eye coordination since the regions were microscopic and the slice wafer-thin. Luca succeeded in capturing the material. But the rest of the fragile slice was destroyed in the process.

As a result, valuable information about how the different minerals connected to one another was lost. To be fair, Luca did not realize at the time how rare and important the specimen would turn out to be, or how essential the information would become. His only goal was to isolate the individual mineral grains as soon as possible so he could take an X-ray diffraction pattern to determine if they were promising quasicrystal candidates.

Once this process was complete, all that remained of the original sample were two tiny specks of material, which Luca glued onto the tips of a pair of slender glass fibers. Tiny as they were, the specks were large enough for Luca to obtain an X-ray powder-diffraction pattern.

Luca compared the results with the published patterns for synthetic quasicrystals and was excited by the possibilities. But he could not be sure if it was a true match. He did not have the computer program needed to perform the sophisticated tests that Peter and I had devised, and he could not detect the rotational symmetry of an atomic arrangement based solely on a powder-diffraction pattern. That was the same problem Peter and I had encountered with the ICDD database.

Within minutes of reading Luca's email, I had forwarded his powder-diffraction data to Peter for immediate study. The test would perform a quantitatively precise comparison of the powder-diffraction pattern of Luca's sample with the data we would expect to see from a natural quasicrystal. Until the results were known, there was no point getting excited by Luca's find.

Two days later, I was back in the United States and received the initial results. Based on the test, there was a reasonably good chance that the grains with the blue dots contained a natural quasicrystal. But it was still too early to get our hopes up. As I explained to Luca, "a good chance" was not nearly the same thing as proof. Peter and I had experienced nothing but false positives during our previous studies. More testing was needed before we could determine if we had actually found a natural quasicrystal.

At that time, however, there were only two tiny specks remaining of the original rock. A quick search for other samples of khatyrkite in collections near Florence or Princeton came up blank. So we had no choice but to focus on the specks we already had in hand. Luca's lab did not have the high-precision instruments needed to perform definitive tests on the remaining material. But I had access to the right equipment as well as the best person to perform the tests. I would call on Nan Yao, director of the Princeton Imaging Center.

On November 11, 2008, about a year and a half since I first started working with Luca, a plastic box arrived at my office from Florence, Italy. The box contained two small brass cylinders used to hold samples for a powder-diffraction experiment. Extending from each cylinder was a thin glass fiber. And glued to the top of each glass fiber was a nearly invisible speck of dark material.

I wanted to make sure the sample had arrived safely. I remember opening the package, taking out the plastic box, and squinting hard to see if I could make out the specks at the end of the fibers. I explained to the student who happened to be in my office at the time that I had spent more than ten years searching for a natural quasicrystal, and was hoping that if I ever found one, it would be at least as large as a pebble.

"It will be awfully frustrating," I said, "if the first natural quasi-crystal turns out to be something that I cannot even see!"

NINE

QUASI-HAPPY NEW YEAR

PRINCETON, NOVEMBER 21, 2008: I kept a firm grip on the small box as I trudged up the hill from my office to the Princeton Imaging Center. Inside were the two brass cylinders I had received from Luca. Each cylinder held a thin glass fiber about an inch long with a precious speck of material glued onto the end.

Nan Yao, the director of the Center, had his head down and was busy working at his desk when I arrived. I took a quick look around his office. Every available nook and cranny was stacked with books, journals, or boxes of samples related to one project or another.

The widespread clutter was tangible proof of the enormous amount of time Nan devoted to working with various faculty members and students from every corner of campus. I myself was already in his debt. He had been contributing some of his own time and discretionary funds to support our investigation.

Nan stood up from his desk and cheerfully maneuvered his tall, slender frame around all the stacks of material to greet me. We exchanged pleasantries, and he invited me to sit down. I looked around, unsure which way to move, because even the chairs and small coffee table in his office were covered with research and debris from earlier meetings. But Nan quickly collected everything and piled it on

the floor in order to make room for me.

Nan, seen on the left in his laboratory, knew that I was bringing him Luca's samples to examine. So I quickly handed him the box and sat back to watch his reaction. Nan is a highly regarded fellow of the Microscopy Society of America, and always manages to maintain a cool and collected professional demeanor. But he was visibly taken aback when he looked inside the box and saw that there were only two specks, each about a tenth of a millimeter in size, to work with. I was already concerned about the lack of material, and Nan's reaction made me even more nervous.

Things are definitely as bad as I suspected, I thought.

Removing the specks from the glass fibers would be a risky endeavor, Nan told me. So before we attempted to do anything that might damage the samples, we decided to see how much we could learn by simply leaving them in place. We would repeat the same X-ray diffraction measurements that Luca had performed but with a more precise instrument to see if the sample was truly as promising as the initial tests suggested.

After several weeks, though, our own results were also inconclusive. Even though Nan was using a better instrument, he could not improve on Luca's results. The powder-diffraction peaks we measured were more or less the same as what Luca had obtained. The flimsy glass fiber mounts were likely the source of the problem, we thought. They wobbled around too much as Nan rotated the sample, smearing the X-ray diffraction pattern.

We considered removing the specks from the original glass fibers and regluing them onto newer, stiffer mounts. But as Nan and I had already discussed, detaching the tiny samples would be a risky operation. If we were going to take such a chance, I decided, we should make it worth our while and not just try to re-glue the samples and repeat the same test. We should jump ahead to the most decisive test: transmission electron diffraction from individual grains in the speck.

The advantage of transmission electron diffraction is that it uses a beam of electrons that can be finely focused. The beam can then be aimed to penetrate a tiny fraction of a single grain of material among the many different grains contained within a single speck. The result is a direct diffraction pattern which reveals the telltale symmetry of the atomic arrangement.

Preparing our sample for the test would prove to be a daunting challenge. It required removing a speck from the glass fiber, separating it into its many microscopic individual grains, and then sorting among all of those grains to find one that would be thin enough for the electron beam to pass through.

Nan's plan was to place a drop of acetone on the tip of the fiber, let the glue slowly soften, and then carefully remove the individual grains of material bit by bit, using a pair of tweezers. His explanation made it sound deceptively simple, but I knew it was excruciatingly detailed work that would require a great deal of skill.

I sat alongside Nan as he brought the dropper to the glass fiber and carefully released a tiny droplet of acetone. I held my breath as the acetone hit the tip of the fiber. Right before our eyes, the entire speck suddenly vanished.

We were both startled. The speck was supposed to contain metallic grains. Metallic grains cannot dissolve in acetone. *What was going on?* Neither of us said anything, but panic and confusion set in. Our eyes slowly shifted downward from the tip of the fiber. And then, seemingly simultaneously, we both let out a gasp.

We had never imagined that the speck was attached to the tip with such a small amount of glue that the tiny drop of acetone Nan had applied would be enough to completely detach it from the mount.

The speck could have fallen onto the floor, where it could have become contaminated. Worse yet, the barely visible speck might have been lost entirely. But as it turned out, about a foot below the glass tip was a table on which a tiny white crucible, about the size and shape of a doll's teacup, had been placed. Nan had put it there to hold bits of material he planned to pick off the speck with his tweezers. Purely by chance, that crucible had been placed directly below the fiber tip.

When Nan and I lowered our gaze, we saw that a droplet of acetone and the entire speck of metallic grains had landed safely right in the middle of the clean white crucible.

The powdery speck of material, which had only been a tenth of a millimeter in size to begin with, had just been split into hundreds of minute grains that were all sitting in a little pool of acetone. We would have to wait for the acetone to evaporate before we could place the particles on a special gold grid, about the size of a small coin, which is commonly used to study powdery samples in a transmission electron microscope.

The electron beam is so fine that only one part of one grain can be studied at a time. The ideal grain would be a pancake shape, wider along two directions but very thin along the direction the beam had to pass through.

As Nan and I looked at the tiny grains in our sample, we realized it would be hopeless to try to slice any of them to the required thickness, which is a thousandth of a millimeter. Our only hope was to find a grain that, by chance, happened to be thin enough and oriented in the right direction for our purposes.

Unfortunately, it would take some time before we could proceed. Princeton's winter break was about to begin, and the Imaging Center would be closed for the holidays. After that, Nan said, the microscope needed for our test was booked solid for the next two months.

I was not optimistic that the sample would prove to contain a quasicrystal. About a decade earlier, Peter Lu and I had studied several minerals with promising powder-diffraction patterns and every single one of them had failed the critical transmission electron microscope test. This sample would probably be no different, I thought. But even so, I did not like the idea of waiting around for months to find out.

I asked Nan if there was any chance that we could check the sample any sooner. He pointed to the only open date on the calendar for the next two months: five o'clock in the morning on Friday, January 2nd. What a surprise, I thought. No one was working the predawn hours the day after New Year's.

But the unfavorable time slot did not discourage me. "Fine!" I said. "See you then!" And, remarkably, Nan agreed.

PRINCETON, JANUARY 2, 2009: When the alarm clock went off at 4:30 a.m., the temperature in Princeton was a frigid 19° F. I bundled up in my warmest winter coat, hat, and gloves and headed off in the dark to meet Nan at the Imaging Lab.

As I drove through town, it suddenly occurred to me that Nan and I had made the appointment weeks earlier, but had never confirmed the date. *Perhaps he forgot?* I wondered. Perhaps I had left my warm bed and was suffering through the freezing cold for no good reason.

But once I arrived at the lab, there was Nan, the consummate professional, working away. I sat down next to him at the transmission electron microscope. He had already carefully placed our grains on a gold mesh sample holder, loaded the holder in the microscope, evacuated the chamber, and was beginning to search among the grains for a promising candidate to study. Nan would be looking directly through the microscope, and I would be able to watch what he was doing by studying the image projected on a nearby monitor.

After a few minutes, he pointed out a grain that was two microns across, about one-thousandth the width of a human hair, seen in the enlarged image below. Under the microscope, the grain seemed to be roughly the shape of a tiny ax. Nan remarked that the part of the grain near the "handle" of the ax appeared thin enough for the electrons to pass through.

He used the controls to slowly move the sample holder until the

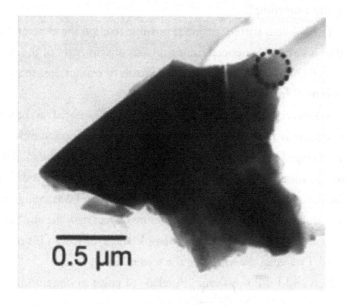

0.5 μm

ax handle was directly in the line of fire of the electron beam. After a few checks, Nan announced he was ready to start the tests.

The first step was to use the transmission electron microscope in "convergent beam mode," which, for highly perfect crystal samples, produces a pattern of crisscrossing ribbons called a "Kikuchi pattern," named for the Japanese physicist Seishi Kikuchi, who discovered the effect in 1928.

To our surprise, the grain immediately produced a beautiful Kikuchi pattern. We did not expect to find such a perfect sample in the rock, and certainly not from the very first spot we tried.

But what really caused our jaws to drop was that the pattern consisted of ten spokes arranged in a ten-fold symmetric pattern, as shown on the facing page. I stared intently at the monitor. A ten-fold symmetric Kikuchi pattern was impossible for ordinary crystals. Finding something like this was our first indication that the sample might actually be a natural quasicrystal.

I felt myself straightening up in my chair. This could turn out to be quite a morning!

The Kikuchi pattern makes it possible to align the electron beam so that it lies nearly perfectly along an axis of symmetry of the atomic arrangement. Nan fiddled with the controls to realign the sample and switch to diffraction mode.

As soon as Nan pressed the switch, an image appeared on the screen that absolutely floored me. I saw a constellation of pinpoint diffraction spots arranged in a snowflake pattern composed of pentagons and decagons, the ideal signature pattern for an icosahedral quasicrystal. I felt a smile coming over my face. I could not believe what I was seeing. It was an electron diffraction pattern much more perfect than the one Shechtman had obtained in 1982. That sample was synthetic. This one was natural. I stared at the image on the screen in awe.

Nan and I did not shout "Eureka!" or cheer or congratulate one

another. In fact, we were dead silent, because no words were necessary. We both knew that we were witnessing a "second kind of impossible" moment. The first-ever discovery of a natural quasicrystal.

Most scientists work for a lifetime hoping to have such a moment. As we sat shivering together in the still frozen lab, Nan and I were keenly aware of how lucky we were. It was a quietly overwhelming moment.

It had been nearly twenty-five years since I first began my informal search for a natural quasicrystal in the mineral collections of natural history museums. And it had been ten years since Nan, Peter, Ken, and I had begun our systematic search of the worldwide database of minerals. Many people thought that laborious project was hopeless and perhaps even a bit silly. And just as the naysayers predicted, we had never once experienced an encouraging result. In fact, we had never even come close.

But that unsuccessful search had led me to Luca Bindi and a

long-forgotten sample in the storage room of his museum. And now, none of the decades of failure meant anything. Nothing made a whit of difference, except what I was looking at on the monitor. Below is a highly overexposed version of the first diffraction pattern we saw that morning.

As Nan and I continued to admire the image, we finally began to exchange a few words. Our conversation was very businesslike as we calmly discussed the next steps.

The first step was to set the sample on a mount that could tilt at certain precise angles in order to observe the different patterns along different directions. This test would be crucial in order to verify that the sample had all the symmetries of an icosahedron.

As Nan explained it, the test could only be done by letting air back into the vacuum chamber in which the sample was held, and then resetting the mount. It was complicated, and would take time to accomplish. Despite the fact that the instrument was officially

booked, the magnitude of our discovery prompted Nan to decide he would try to steal some time later in the week to perform the test. For now, it was time to go home.

The freezing-cold streets of Princeton were still dark and deserted as I headed out of the laboratory. But I took no notice of the frigid temperature. I drove home almost in a dream state, replaying the morning over and over again in my mind. *A natural quasicrystal. Impossible.*

A few hours later, after a brief nap, I sent an email to Luca with the subject line: *Quasi-Happy New Year.* My Italian colleague was the third person in the world to know that natural quasicrystals had just been discovered. Luca, though, would probably say that he was the first. The initial powder-diffraction tests at his Florence lab had been inconclusive. But Luca's scientific intuition, which would prove to be remarkably accurate over the next few years, had given him complete confidence that the material he sent me contained a quasicrystal.

A few days later, Nan was able to steal a bit more time on the transmission electron microscope, as promised. He rotated the sample at various angles and discovered a series of diffraction patterns with the symmetry of a rectangle (left image below) and a hexagon (right image below).

The angles by which Nan needed to rotate the sample to go from

the ten-fold pattern to the rectangular and then to the hexagonal, turned out to correspond to precisely the angles predicted for an icosahedron; for example, the angle between an imaginary line that passes between the center of an icosahedron and one of its corners and a line that passes between the center of the icosahedron and the center of one of its triangular faces. It was indisputable proof that our grain had perfect icosahedral symmetry.

Luca's preliminary test indicated that the grains contained aluminum, iron, and copper in roughly the same proportion that An-Pang Tsai and his collaborators had measured in the historic sample they discovered in 1987, which was the first bona fide example of a synthetic quasicrystal with pinpoint diffraction peaks. But we needed to make a more accurate measurement to be certain.

I had a small sample of Tsai's synthetic quasicrystal, which had been given to me as a memento in 1989. The nugget was a prized possession displayed in my office for more than twenty years, but I broke off a small chip for Nan so that he could make a quantitative comparison between the synthetic sample and the first known natural one.

The match was nearly perfect: $Al_{63}Cu_{24}Fe_{13}$ (63 percent aluminum, 24 percent copper, and 13 percent iron). Tsai's beautifully faceted, dodecahedron-shaped synthetic quasicrystal and the tiny grains in the natural khatyrkite sample had precisely the same atomic arrangement and composition.

These two substances had arrived in Princeton from opposite ends of the world. One of them was manufactured in a Japanese laboratory, the other created by nature and imported from Italy. And now, the two of them were discovered to be nearly a perfect match. *Impossible.*

Luca and I drafted the paper, entitled "Discovery of Natural Quasicrystals," with help from Nan Yao and Peter Lu, which we submitted to *Science* magazine, a leading journal for presenting new results. I knew we would have to wait several months to see if the article was accepted for publication.

At this point, I should have been celebrating that we had finally succeeded in finding a natural quasicrystal, a goal that I had been pursuing for decades. But, instead, I felt strangely dissatisfied. I had a nagging sense that nature was still hiding something about the khatyrkite sample that had yet to be discovered.

I could not put my finger on exactly what it was that made me feel that way, and I had no idea how long it would take before I would find out. There was just an overwhelming sense that the adventure was just beginning.

TEN

WHEN YOU SAY IMPOSSIBLE

PRINCETON, JANUARY 8, 2009: I knocked sharply on the tall oak door with a glass insert labeled "PROF. L. HOLLISTER." It was to be the first of many meetings with the renowned geologist, and I had no idea what to expect.

I knew Lincoln was an expert in petrology, the study of the origin and the composition of rocks. I also knew he was a hard-nosed scientist with a wide-ranging set of interests and was beloved on the Princeton campus. What I did not know was that he was about to become one of our strongest critics and would soon be questioning the validity of our entire project.

Lincoln had made a career of challenging convention and ultimately proving his case. When he first entered the field in the 1960s, the standard view was that minerals in metamorphic rocks have uniform compositions because they formed at high temperatures and pressures. But Lincoln was able to explain why this is not the case. As one of the first geologists to receive rocks brought back from the moon, he showed that certain minerals in lunar lavas believed to have come from deep below the moon's surface at high pressure had, in fact, developed in rapidly cooled lavas at the surface. He also significantly advanced our understanding of the continental crust with a series of explorations of remote regions of British Columbia, Alaska, and Bhutan.

Throughout his career, Lincoln's success relied on his survival skills in the wild and a tough, no-nonsense approach in the laboratory. I knew there was no better expert to help me figure out how our natural quasicrystal had formed.

Man-made quasicrystals had now become commonplace all over the world, ever since An-Pang Tsai had fabricated the first perfect sample of $Al_{63}Cu_{24}Fe_{13}$ in 1987. But Tsai had worked under carefully controlled laboratory conditions, beginning with just the right proportion of different metals and carefully regulating the rate at which he cooled the mixture. As a result, his team produced perfect man-made samples like the one on the left below. By contrast, the quasicrystals we discovered in the Florence sample were formed by nature in a totally uncontrolled environment and squeezed together with other minerals in a hodgepodge, as shown in the image on the right below. The white dots correspond to the locations of quasicrystals, but the other shaded dots correspond to various other crystalline minerals.

Nature's quasicrystal had the same atomic makeup as a synthetic quasicrystal, and they both had approximately the same defect-free

structure. It was as if we were looking at identical twins born of different parents from distant parts of the world. *How did that happen?* I wanted to know.

When Lincoln opened the door to welcome me, my first impression was: *This fellow really looks the part of a geologist.* Standing 5-foot-10 with a tanned complexion, silvery hair, and rugged good looks, Lincoln appeared ready to pick up a backpack and head off at a moment's notice for another one of his outdoor adventures.

In fact, he looked so physically fit that, had he not mentioned it, I would never have guessed that Lincoln was seventy years old and on the verge of retirement. He was packing up his office, he said, which explained why everything was in a state of disarray with maps, microscopes, and large rock samples strewn everywhere.

Lincoln invited me into his inner office where there was more room to sit down, and I spent the next thirty minutes explaining my story. I

told Lincoln about how we had first developed the theory of quasicrystals, the discoveries of synthetic versions in the lab, and my search for a natural sample that had begun in the 1980s and culminated less than a week earlier with the discovery at Princeton's Imaging Center.

And then I asked him a question that had been bugging me ever since: *How had nature done it?*

Lincoln narrowed his eyes and glared at me. I

learned afterward that his students are all too familiar with that look. They call it the "Hollister gaze," and it invariably means trouble.

He must have felt sympathy for the fact that I was a theoretical physicist who obviously knew virtually nothing about geology. The Hollister gaze slowly softened as he began to break the bad news to me gently.

"What you have there is . . ." he said, allowing a long, dramatic pause, ". . . impossible!"

"Wait," I quickly interrupted before he continued. I had been hearing that word for decades and wanted a chance to explain.

"Quasicrystals are definitely possible," I reminded him. "We have fabricated them in the laboratory, including ones with the very same composition as the natural sample that we have just discovered."

Clearly struggling to be patient with me, Lincoln raised his voice a notch or two. "I am not concerned about the quasicrystal part," he said forcefully. "I never heard of them, but what you have explained sounds okay. What concerns me is that you said that the quasicrystal and the crystal khatyrkite both include metallic aluminum.

"Aluminum has a very strong affinity for oxygen," he stated. "There is plenty of aluminum on the Earth, but it is not metallic. It is all bonded to oxygen." Once aluminum bonds to oxygen, it is no longer shiny and does not easily conduct electrons like metallic aluminum.

"As far as I know, no sample of metallic aluminum or any alloy containing aluminum metal has ever been seen in nature. You think you have a natural rock. But, I am sorry to tell you, it is probably refuse from an aluminum smelter." The metallic aluminum encountered in everyday life is all made synthetically by separating metallic aluminum from aluminum oxide.

Lincoln's pronouncement sounded definitive. At this point, most geologists, respecting his reputation and hearing his stern tone of voice and all it conveyed, would have thanked him for his advice and ended their investigation right there and then.

Sitting before him, though, was a stubborn theoretical physicist who was a neophyte when it came to geology but all too familiar with impossible challenges. So I persisted and asked Lincoln the same question I always ask myself whenever I hear the word "impossible."

"When you say 'impossible' . . . Do you mean impossible like 1 + 1 = 3? Or do you mean very, very unlikely? And if true, very, very interesting?"

Thankfully, Lincoln did not seem to think my question was impertinent because he did not immediately throw me out of his office. Instead, he paused a few moments to consider the question. When he finally spoke, his voice had returned to a more normal volume.

"I suppose," he said thoughtfully, "that if I were forced to come up with a natural explanation, I would need to find conditions where aluminum could easily separate from oxygen. It would require ultra-high pressures, which can be found three thousand kilometers below the surface of the Earth near the core-mantle boundary.

"But then," he continued to speculate, "if you could manage to make metallic aluminum and form your quasicrystal, you would need to find a mechanism to get it to the surface of the Earth rapidly without the mineral decomposing and the aluminum reacting with oxygen on the way up."

For a moment, I was worried that he might consider this an impassable roadblock. Happily, I was wrong.

"There is a conceivable way this can happen," he offered. "You may know of Jason Morgan, the Princeton geoscientist who helped to establish the modern theory of plate tectonics.

"Jason retired a few years ago. He had a theory that there might be superplumes, tube-like upwellings of material from the core-mantle boundary to the surface. The superplumes, if they exist, would be giant versions of the well-known plumes that created the Hawaiian Islands.

"The superplume idea has never been proven," Lincoln continued. "But if your sample was made at the core-mantle boundary and carried to the surface in this way, it would be the first direct evidence of his idea."

When Lincoln finished, my eyes must have been the size of saucers. *Our sample is not impossible after all*, I thought. *And, if it turns out to be of natural origin, it is going to be extremely important.*

After a brief silence, I timidly offered my pet idea. "If the problem is keeping aluminum away from oxygen, is it possible that the sample was made in space, perhaps inside a meteorite?"

The idea that meteorites might be a source of quasicrystals had occurred to me before. I had been thinking about it for several years and had even mentioned the possibility to Luca, although we never pursued it. I was not aware at the time, though, that my question was terribly naive. I thought there was little or no oxygen in space when, in fact, meteoroids and asteroids are full of oxygen bonded to other elements.

Fortunately, Lincoln did not point out the error in my thinking. "I do not know much about meteorites," he said, "but I know someone who does."

Lincoln was referring to Glenn MacPherson, the head of the Division of Meteorites at the Smithsonian National Museum of Natural History. Glenn received his PhD from Princeton in 1981. Lincoln had known him for decades and had even recommended him for his current position.

Lincoln offered to arrange a visit for me to meet Glenn at his office in Washington, D.C. He also offered to accompany me on the trip, and I eagerly accepted. I interpreted the offer to be a positive sign that the legendary geologist was somewhat excited by our finding.

Once I returned to my office, I emailed Luca in Italy to tell him about my meeting with Lincoln. Luca was aware of Lincoln's professional reputation and held him in high esteem. I tried to be as upbeat as possible and did not dwell on the fact that Lincoln's first impression was that the sample was just an ordinary piece of scrap metal.

Luca, like me, was not aware that metallic aluminum had never been found before in nature. That bit of news meant we had even more reason to worry about the reaction to the *Science* article we had submitted. Not only were we reporting an impossible new form of matter, a quasicrystal, we were also announcing that we had found natural metallic aluminum, making our discovery doubly impossible.

Luca was impressed with Lincoln's superplume idea and at how well the conversation seemed to have gone overall. But in truth, I was becoming concerned. Both of the proposed explanations, superplumes and meteorites, sounded like long shots.

A week later, Luca and I received good news from the editors of *Science* magazine. The manuscript describing our discovery of the first natural quasicrystal had passed the first round of review. This was promising. The fact that the editors had not rejected our paper outright meant they did not consider the evidence to be ridiculous, even though the quasicrystal included metallic aluminum. The real test, though, would be the next round of scientists reviewing the article. They would all be experts in the field who, much like Lincoln, would probably consider the reported discovery of natural metallic aluminum to be preposterous.

WASHINGTON, D.C., JANUARY 24, 2009: "Impossible!"

As Lincoln and I climbed the steps to the entrance of the Smithsonian, Glenn MacPherson stood at the top of the stairs. He was holding open the giant door to the Natural History Museum while blurting out his opinion of the Florence sample loud enough for everyone to hear.

I did not know that Lincoln had prepped Glenn about the subject of our meeting and that, by the time we arrived, he had already given our sample careful consideration. So I was completely startled by the way Glenn introduced himself.

Glenn was taller than Lincoln and me. He was slender with dark hair graying at the temples, had a dark mustache and, unlike Lincoln, the look of someone who spent all of his time in the laboratory.

Glenn ushered Lincoln and me inside and helped us register for the special badges we would need to enter the inner sanctum of the Smithsonian. He then proceeded to lead us on a long, labyrinthine path to his office that included countless corridors, elevators, a slew of security doors, and then, yet again, even *more* corridors. The entire time we followed him through all of those endless passageways, Glenn continued to pummel me with all of the reasons our sample could not possibly be natural.

The presence of metallic aluminum, the same concern that Lincoln had raised, was only the first of the problems, according to Glenn. Once we finally arrived at a meeting room near his office, he asked us sit down at a big table where he proceeded to lay out a series of key papers and data demonstrating how impossible it is to form metallic aluminum naturally on Earth.

"And as for meteorites . . ." Glenn said ominously, as he began to take aim at my pet theory. In all of his experience with meteorites, and Glenn had seen all types, he assured us, he had never seen a single example that contained any metallic aluminum or aluminum alloys.

Glenn was convinced, absolutely convinced, that our sample was . . . and then he uttered the dreaded four-letter word . . . S-L-A-G. Slag is a catch-all name for a useless by-product of an industrial process. Slag meant unnatural. Slag meant we had failed to find what I thought we had found. Slag was the ugly word I did not want to hear.

Glenn was not finished with his case, however. Another big

problem, he explained, was our claim that the metallic aluminum was supposedly mixed with metallic copper in three different minerals in our sample: khatyrkite, cupalite, and our quasicrystal. That is *also* impossible, he asserted. Just as aluminum has an affinity for oxygen, copper has an affinity for sulfur.

The two metals are found in different classes of minerals because of their different chemical bonding patterns. It would be inconceivable, according to Glenn, that they could naturally form a metallic alloy like khatyrkite, cupalite, or our quasicrystal through any natural geochemical process.

A third issue was the absence of any corrosion. How could a sample containing aluminum metal survive on the surface of the Earth without any sign of rust?

Glenn then continued to rattle off a seemingly endless list of additional reasons why the sample could not possibly be natural.

As I listened and took notes, I could tell that Glenn had put considerable thought into the issue and presumed that he was trying to impress Lincoln, his former mentor. At first, Lincoln tried to defend the case for a natural quasicrystal by presenting his novel idea about superplumes. But he eventually wilted under Glenn's continued barrage. By the time we left the Smithsonian a few hours later, Lincoln seemed completely convinced by Glenn's conclusion: Our sample had to be an artificial by-product of an aluminum smelter or laboratory.

The Smithsonian meeting could have marked the end of the investigation. I was sure that Lincoln and Glenn thought they would never hear from either me or Luca again.

But I was not swayed by Glenn's arguments. They all relied on a variety of reasonable but, truth be told, unproven scientific assumptions. And while Glenn had presented abundant evidence supporting his case, all of his evidence was, by definition, a reflection of what had been observed in the past. None of it proved that it was impossible to find something new in the future.

I preferred to view the situation another way. If Glenn was wrong and the Florence sample was not slag, then it represented something even more spectacular than we first imagined. It would not only serve to prove the existence of natural quasicrystals, it would also overturn widely accepted assumptions about the kinds of minerals that can form in nature.

Once Lincoln and I returned to Princeton, I wrote an email to Luca containing a full and brutally honest summary of the visit. I hit the send button, and wondered if the disappointing news would make him decide to abandon the project. I did not have to wait very long for an answer. Within minutes, a reply appeared in my mailbox.

Luca had no intention of giving up. He was confident that our sample was natural. Not only that, he was every bit as committed to the investigation as I was and equally determined to work with me to prove the case scientifically. Luca and I acknowledged to each other that we were charting a dangerous course. It would be a very public battle that could, despite our best efforts, prove to be very embarrassing.

In order to proceed, we would need to formulate a new strategy. And we would need our two harshest critics, Lincoln and Glenn, to play a key role.

ELEVEN

BLUE TEAM vs. RED TEAM

PRINCETON AND FLORENCE, JANUARY 25, 2009: Luca and I were under enormous pressure. We had written and submitted the scientific paper announcing our discovery and the review process was well under way. But now we were getting major pushback from Lincoln and Glenn, neither of whom agreed with our conclusions.

The sample was slag, they argued. We had been hoodwinked. Natural khatyrkite, our natural quasicrystal, and metallic aluminum were most certainly impossible.

Their opposition put us in a terrible bind. If the *Science* paper was published and later turned out to be wrong, as Lincoln and Glenn believed, the damage to our reputations and impact on future research projects would be enormous. On the other hand, if we pulled back and withdrew the paper, the reversal would draw attention and raise suspicions. The search for natural quasicrystals would lose credibility in the scientific community and perhaps come to an end altogether.

The only way out of the predicament was to do our best to work quickly to resolve the central scientific issue before publication. *Is the quasicrystal natural or slag?* In our view, the evidence strongly tilted toward the sample being natural. But we needed more than that. We needed substantial evidence, enough to sway our harshest critics.

It was essential to keep Lincoln and Glenn involved in the

investigation. For one thing, the four of us made a good team. Their scientific expertise was complementary to ours. The fact that they were both extremely skeptical was an advantage, I thought. No matter how hard we might try, I did not believe Luca and I could trust ourselves to be completely objective.

> The first principle is that you must not fool yourself and you are the easiest person to fool.
>
> —Richard Feynman, "Cargo Cult Science," 1974

My early mentor, Richard Feynman, delivered an elegant speech at my Caltech commencement ceremony about the dangers of a phenomenon known as confirmation bias. It is a well-known human frailty that has been studied for decades. People in every walk of life tend to ignore evidence that runs contrary to their preexisting beliefs and eagerly accept evidence that appears to support them. Feynman's message was that the more you believe in something, the more vulnerable you are to making a mistake.

I have always strongly embraced this philosophy. Over time, I developed a tried-and-true solution to the problem—I always seek out people for my research teams whose role, in my mind, is to be the fiercest critic imaginable. My in-house critics must be harsher than anyone else who might challenge the work if it were ever published. I assign critics to the "red team" and advocates to the "blue team." The goal is for the two teams to duke it out in a merciless, but friendly, competition until the scientific truth is revealed.

Lincoln and Glenn were both so negative at this point that they were perfect candidates for the red team. That was to be their implicit role, even though we never discussed it directly. Luca and I would represent the blue team and be primarily responsible for gathering the probative evidence.

The blue team advocates, Luca and I, immediately began holding daily meetings over the Internet to discuss our research efforts, which quickly became something of a roller-coaster ride. A thrill at one moment and a fright the next. Before too long, we found ourselves addicted to the adrenaline rush.

Luca suggested we engage in typewritten chats instead of verbal conversations, which was remarkably prescient. The written records would turn out to be a valuable resource during the tortuous course of our investigation. We often went back to them to check facts and refresh our memories.

Our intense daily chats inevitably turned into a rivalry: Which one of us could discover the most interesting item? We competed to find the best new scientific paper, best new Internet contact, best new hint about the origin of the Florence sample, and best new laboratory measurements of the remaining specks. On most days, Luca was the clear winner. But I enjoyed an upset victory every now and then.

Our first priority was to figure out how and when the sample labeled "Khatyrkite" made its way to Luca's mineral museum.

Luca combed the museum archives and dug up correspondence dating back more than two decades. The letters revealed that his museum acquired the khatyrkite in 1990 as part of a larger purchase of 3,500 specimens. Curzio Cipriani, Luca's predecessor, paid roughly $30,000 for the entire lot. It was amusing to learn that our now-precious khatyrkite once had a whopping street value of less than ten dollars.

According to the records, Cipriani had purchased the specimens from a private mineral collector in Amsterdam named Nico Koekkoek. It was promising information, but woefully incomplete. None of the old paperwork included any contact information.

AMSTERDAM, HOLLAND, FEBRUARY 2009: Luca and I jumped on-line and started plowing through Dutch telephone directories. We spotted numerous Koekkoeks, but no one named Nico. We found a number of other mineral dealers and bombarded them with emails, both in English and Dutch, pleading for help. Despite a month of concerted effort, we failed to unearth a single clue.

Without Nico Koekkoek, I wondered, *how could we ever hope to establish the origin of the Florence sample?*

It was a disappointing dead end, but Luca and I were already deeply engrossed in other aspects of the investigation. Time was so short that we had no choice but to pursue many different ideas simultaneously.

One of our biggest headaches was that Lincoln and Glenn kept insisting that the aluminum-containing metal alloys in the Florence sample were nothing but slag. Granted, khatyrkite and cupalite were listed in the International Mineralogical Association catalog of rec-ognized minerals. But neither Lincoln nor Glenn trusted the analysis connected to those entries. Metallic aluminum without oxygen was the sticking point. Impossible, they both said dismissively.

Luca and I thought we could persuade them that the metal alloys were natural by finding another sample of khatyrkite in a different collection. We knew the source would have to be unimpeachable in order to convince them.

We began by checking prestigious museums with gigantic min-eral collections, like the Smithsonian in Washington, D.C., and the American Museum of Natural History in New York City. No luck in either place, which was, frankly, unexpected and a bit concerning. Then we turned to museums with more modest collections, some of which had mineral catalogs we could review online. Once again, no luck, which was now becoming a bit more worrisome. We then began checking lesser collections in smaller museums, academic institutions, and individual collections around the world.

We reached out to international mineral dealers. *Did any of them have khatyrkite, or had they ever sold khatyrkite to anyone?* We checked Mindat.org, an excellent public database of minerals used by amateur and professional mineralogists. *Did anyone on that website have any khatyrkite?*

NORTHFIELD, MINNESOTA, MARCH 2009: The upshot of our exhaustive worldwide search was a grand total of four potential sources of khatyrkite. Three of the samples were in North America and Western Europe. A fourth, perhaps the most promising, was in St. Petersburg, Russia.

I was especially excited when Luca discovered that one of the samples was held in the mineral collection at Carleton College in Northfield, Minnesota. *A sample kept at an academic institution is certain to be authentic*, I thought. I was even more confident when I learned that Carleton's leading geology professor, Cameron Davidson, was a Princeton graduate and one of Lincoln's former students.

Cameron agreed to send me the mineral for examination. I had high expectations for that particular sample, and began anxiously checking my university mailbox several times a day. It was more than a week before the bad news arrived—Cameron had decided to test the mineral himself and found it to be a complete fake. It was labeled "Khatyrkite, a metallic alloy of aluminum and copper." But testing proved there was no trace of aluminum whatsoever.

Similar things happened with two of the other supposed samples identified in our search. In the end, testing revealed that every sample outside of Russia was an absolute fake.

Our experience trying to chase down a fresh sample of khatyrkite revealed the limitations of the international mineral market. Amateur collectors are eager to get their hands on as many different types of minerals as possible. But there is no way to authenticate a mineral with the naked eye. Unlike diamonds, which are so costly that independent certifications are considered routine, most minerals are modestly priced and professional tests are time-consuming and comparatively expensive. So more often than not, an amateur collector is willing to purchase a sample based solely on a dealer's representation. But most likely, the dealer has not performed any testing.

Eventually, a collector may decide to either donate or sell an untested mineral to a museum or academic institution. The curator is then in the same dilemma as the collector. Testing is time-consuming and comparatively expensive. A common approach is to simply accept the stated designation.

All of the fake samples had proven the same thing: The international mineral market was like a big casino, and every mineral sale a toss of the dice. I was beginning to appreciate why Lincoln and Glenn were so skeptical about the Florence sample, even though it had been found in a respected museum.

Maybe it was not legitimate after all?

St. Petersburg, Russia, February–March 2009: We were down to our fourth and final prospect, which was at Russia's St. Petersburg Mining Museum. I tried to temper my enthusiasm in light of our previous failures, but was still betting on success.

The Russian sample has to be legitimate, I thought, *because it is the official "holotype" for crystal khatyrkite, which presumably means it has been rigorously authenticated.*

A holotype is the certified example of a new mineral, as approved by the International Mineralogical Association. In order to have a new mineral accepted by the IMA, one must submit results of a series of laboratory tests that are reviewed by an international committee of mineralogists. If the committee finds the tests convincing, a paper describing the new mineral must be submitted for publication. In addition, a sample, called a holotype, must be donated to a public museum.

There were three Russian scientists connected to the khatyrkite holotype—Leonid Razin, Nikolai Rudashevsky, and Leonid Vyal'sov. Luca and I knew they had coauthored a 1985 scientific paper reporting the discovery of both khatyrkite and cupalite, as shown below. That was notable for us, because it hinted at a match. They were the same rare minerals we had found in the Florence sample.

Luca and I believed that if the St. Petersburg holotype was authentic, as one would expect, the argument in favor of the Florence sample would be greatly strengthened. So we returned to the Russian paper and reexamined it in earnest.

The new mineral was described in the paper as being found in the

ЗАПИСКИ ВСЕСОЮЗНОГО МИНЕРАЛОГИЧЕСКОГО ОБЩЕСТВА

Ч. CXIV   1985   Вып. 1

НОВЫЕ МИНЕРАЛЫ

УДК 549.3 (571.6)

Д. члены Л. В. РАЗИН, Н. С. РУДАШЕВСКИЙ, Л. Н. ВЯЛЬСОВ

НОВЫЕ ПРИРОДНЫЕ ИНТЕРМЕТАЛЛИЧЕСКИЕ СОЕДИНЕНИЯ АЛЮМИНИЯ, МЕДИ И ЦИНКА — ХАТЫРКИТ $CuAl_2$, КУПАЛИТ $CuAl$ И АЛЮМИНИДЫ ЦИНКА — ИЗ ГИПЕРБАЗИТОВ ДУНИТ-ГАРЦБУРГИТОВОЙ ФОРМАЦИИ [1]

Среди природных образований впервые обнаружены соединения алюминия с медью и цинком. Они находятся в тесном срастании и представлены мелкими (размером от долей до 1.5 мм) неправильной формы, угловатыми стально-серовато-желтыми металлическими частицами, внешне схожими с самородной платиной. Эти частицы встречены в черном шлихе.

"Koryak-Kamchatka" region, across the Bering Strait from Alaska. The Kamchatka Peninsula is an arc of land created by active volcanoes that lies between the Sea of Okhotsk on the west and the Pacific Ocean on the east. Based on the paper, we often referred to the locality as "Kamchatka" in our discussions, presentations, and papers, and "Kamchatka" will often be used in this book as well to refer to the discovery site.

In actuality, though, the location is in the Koryak mountain range, an area just north of the Kamchatka Peninsula that is part of the Chukotka Okrug, as shown on the top panel on the next page. The word "okrug" denotes an administrative district in Russia.

One of the Koryak's largest drainage rivers is the Khatyrka, from which the name "khatyrkite" is derived. According to the Russian scientists, they discovered khatyrkite while panning blue-green clay near the Khatyrka River along the Listvenitovyi Stream, as shown on the bottom panel of the next page.

Luca and I were especially excited to discover that the location matched the label on the plastic box Luca had found in his museum: "Khatyrka, Koriak Russia."

Since the label matched, did that mean the Florence sample was from the same place? Perhaps. If so, it would seem likely that the sample was nature-made, since there were no Russian foundries or factories operating in that remote region.

But even if they were not from the same place, the mere existence of an authentic sample of khatyrkite with the same basic chemical composition as the one found in Florence was good news for the blue team. If we could prove to Lincoln and Glenn that the St. Petersburg holotype was of natural origin, that would force them to reevaluate their opposition.

It was clear what we needed to do next. We needed access to the holotype in order to verify the original results of the laboratory tests.

Luca and I tried to use our combined influence to borrow the

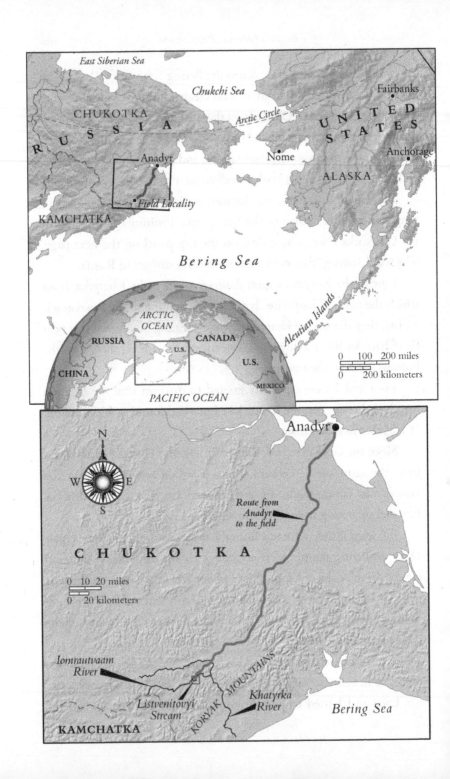

holotype from the Russian museum. We explained that we were trying to authenticate the Florence sample and wanted to conduct a special series of noninvasive tests on the holotype that would not damage the material in any way. Unfortunately, the director of the St. Petersburg Mining Museum refused to cooperate. It is quite common for scientists to lend each other samples for testing, especially if the original sample will not be damaged in the process, as would have been the case here. But the Russian director strictly prohibited anyone from touching the holotype, including his own team of in-house scientists.

This was an especially difficult defeat for Luca and me to accept. Another disheartening dead end.

PRINCETON, MARCH 2009: In the midst of all this activity, *Science* magazine sent another update about our paper announcing the discovery of a natural quasicrystal. I had been dreading this moment for months. Luca and I were currently bearing the full brunt of Lincoln's and Glenn's harsh criticism, and I assumed their opinions would be shared by other geologists. So I was expecting a round of withering criticism from the magazine along with a stinging rejection letter.

Having braced for the worst, I was pleasantly surprised and gratified to read the reviews and analyses of our paper. The anonymous team of professional peers was generally supportive. They recognized the significance of the discovery, posed good questions, and offered constructive suggestions for improvement.

Luca and I did not have any difficulty incorporating the referees' suggestions. Both of us felt the paper was likely to be accepted once our revisions were considered, which meant publication might now be less than two months away. That was our desired outcome, of course. But it created an even greater sense of urgency to resolve our differences with the red team.

Lincoln suggested a novel approach. If we could establish the precise location where the St. Petersburg sample was discovered we could study geological conditions in the vicinity. We might find something, he speculated, that would help explain the baffling presence of metallic aluminum.

Lincoln and I immediately went to work in Princeton's Maps and Geospatial Information Center. We spent hours poring over maps of Chukotka searching for the slightest mention of the Listvenitovyi Stream in the large maps held in the collection. This was time-consuming, old-school research.

The Russian team's paper included enough information for us to narrow the search to within ten or twenty miles. Normally, that would be helpful. In this case, though, it left us with too broad an area to evaluate. Terrain in the Koryak mountain range is so varied that geological conditions change dramatically every few miles. We needed to find the precise stream in order to get as close as possible to the discovery site.

I always found the name of the stream to be musical, at least the way I pronounced it: *LIST-ven-i-TOV-yi*. I repeated the name over and over to myself while scanning the map pages, as if I could conjure it up by silently chanting its name. *LIST-ven-i-TOV-yi. LIST-ven-i-TOV-yi. LIST-ven-i-TOV-yi.* Perhaps that was how it managed to burrow its way into my subconscious.

I seldom remember any of my dreams. But after returning home late one night from the map library, I had a particularly vivid dream involving the Listvenitovyi. Luca and I were standing together along the side of the stream in front of a hill that rose several feet above our heads. Our hands were clasped together and held high as a sign of victory, and we were smiling broadly.

I could never imagine myself traveling to a place as remote as the Koryak Mountains. But the dream was such an intense emotional experience that I wrote it down and mentioned it to Luca during one of our daily Internet chats. It was such an unusual experience, I could only surmise that all of the failures and frustrations of the blue team were beginning to take a psychological toll on me.

Despite the long hours we spent searching every available resource, Lincoln and I were never able to find any trace of the Listvenitovyi Stream on a map. So once again, another dead end.

The blue team vs. red team competition was becoming so one-sided that I was no longer keeping score.

PRINCETON AND FLORENCE, MARCH–APRIL 2009: Even though most of our efforts were focused on trying to establish the origin and legitimacy of the St. Petersburg sample, the blue team's investigation was also continuing on a number of other fronts. For example, Luca and I were still struggling to find a scientific explanation that could account for the presence of natural aluminum.

We were surprised to discover that there were already a number of papers by scientists claiming to have discovered pure metallic aluminum in nature. Not mixed with copper or other metals, like our sample, just pure aluminum. When Luca and I showed the papers to Lincoln and Glenn, they scoffed at every claim. The authors were not well-known, they said, and the evidence not convincing. Natural aluminum was definitely impossible, as far as the red team was concerned.

Nevertheless, I contacted the scientists involved and began purchasing samples of their material, which is how I ended up acquiring what I like to call the "World's Largest Collection of (Purported) Natural Aluminum."

Once I started examining the samples, I had to admit that Lincoln and Glenn might be right. Most of them looked questionable.

One of them was especially suspicious. It looked like a fragment of electrical wire that had been struck by lightning. Others were harder to assess, and I thought a more serious analysis was justified. I could have tested them myself, but decided they should be evaluated by the person most skeptical of their validity.

So I took my entire collection to Glenn, hoping he could examine everything in his lab at the Smithsonian Natural History Museum. But somehow, he just never found the time. Perhaps he was too skeptical or too busy or both. As of this writing, my collection is still awaiting his review. It is not taking up too much room on his shelf. The "World's Largest Collection of (Purported) Natural Aluminum" could fit in the palm of your hand, with room to spare.

Luca and I found numerous other papers describing metallic aluminum found in remote regions that could definitely be traced back to human activity. Sources included foundries, jet fuel exhaust, nuclear bomb tests, frying pans, and coins left on hot stoves for long periods of time. Based on our study, all of the samples produced by these anthropogenic processes had physical properties that we did not find in the Florence sample. That did not prove our sample was natural. But it was a bit of good news that briefly buoyed the lagging spirits of the blue team.

Luca and I also found scientific papers suggesting various theories speculating how metallic aluminum might form as part of a natural process. Some of those ideas seemed convoluted. We didn't know how to judge their viability.

I had originally hoped that we would be able to screen out most of these theories based on the geology of the Kamchatka region. Unfortunately, that turned out to be a naive idea. Whatever crazy geological property was needed to support a theory about the formation of natural aluminum, Chukotka had it. The region was literally a geological mash-up, which explained why geologists had been studying it for decades. Because of the geological complexity of the region, not a single one of the theories could be eliminated.

TEL AVIV, ISRAEL, MARCH 2009: The deadline to decide whether to publish our *Science* magazine paper was fast approaching. But so far, Luca and I had failed to prove that our quasicrystal discovery was natural, as we believed it to be. So we made a last-ditch effort to find Leonid Razin, the lead author of the 1985 Russian paper that had first reported the discovery of khatyrkite and cupalite in Chukotka.

Following the limited amount of information available on the Internet, we learned that Razin was head of the Soviet Institute of Platinum in 1985, when the paper was published. That made perfect sense to us, because platinum has strategic technological applications, and there are significant deposits of platinum in Chukotka. Razin's professional appointment helped explain his presence at the remote location.

The fact that Razin was head of the Institute of Platinum suggested to us that he was no ordinary mineralogist. He would have needed important political connections within the Communist Party to be named to that post.

Was Razin still alive? Was he still in Russia? Luca and I emailed Russian scientists. Each of them referred us to others, who referred us to others, who referred us to others. Down the rabbit hole we went, once again.

We eventually learned that Razin was a well-known character, but never particularly popular or admired by colleagues. Several people told us he had powerful KGB connections and wasn't afraid to use those connections to destroy his competitors.

Others, including internationally known geologists and members of the prestigious Russian Academy of Sciences, told us that Razin was not trustworthy. They did not believe Razin's claim of finding natural minerals containing metallic aluminum solely because they did not consider him to be a reliable source about anything. In other words, our Russian colleagues agreed with Lincoln and Glenn, but for different reasons. They, too, thought the sample was probably a fake.

That was frankly the last thing Luca and I wanted to hear. At this point, Razin's paper was our only active lead. We were still hoping the Russian discovery would turn out to be legitimate, even though all the evidence was stacking up against us.

After multiple email chains, we finally found several people who told us Razin was still alive. They believed he had emigrated from Russia to Israel sometime after the fall of the Soviet Union in the early 1990s. Israel is not a very big country, so it was a relatively easy process to review the online telephone books for each metropolitan area. I quickly found an "L. Razin" listed in Tel Aviv.

I tried calling the number. Problem: Whoever answered the phone could not speak English. I hung up, and dragooned an Israeli graduate student at Princeton to act as translator.

I called again with my expert Hebrew translator in tow. Problem: The residents did not speak Hebrew, either.

The third time around, I drafted a Russian graduate student to help. Success at last. The residents spoke fluent Russian and immediately confirmed that I had reached the home of Leonid Razin.

I took several deep breaths while waiting for Razin to come to the phone. I realized that we were about to have a momentous conversation that would affect the future of the entire project.

After a brief introduction, I told him I was interested in a 1985 paper about the discovery of khatyrkite and cupalite.

"Are you the Leonid Razin who is the lead author on the paper?" I asked, trying to contain my excitement.

"Yes, I am Academician Razin," he answered coldly.

Razin sounded more formal than friendly. He obviously wanted to make sure I recognized his status as a distinguished member of the Russian Academy of Sciences.

I decided not to tell him that I had comparable status in the United States and had been elected to the National Academy of Sciences. Instead, I tried to put him at ease by praising his paper and describing how we had discovered samples of a new phase of matter in a rock with a similar chemical composition.

His reaction was surprisingly tepid. I had expected that Razin would be excited that another scientist was calling to discuss a paper he had written nearly a quarter century ago. I had anticipated that he would be even more excited to learn that his work might help establish the foundation for a new form of matter.

But instead, Razin seemed to exude extreme indifference. I found his attitude off-putting, but continued to pepper him with questions.

"Were you the one who personally discovered the khatyrkite samples in the field?"

"Da," he responded. I could understand that answer without any translation, and smiled with relief.

"Do you have your geological field notebook?" I asked, hoping to be able to read about how he had discovered the khatyrkite sample along with his notes on the geological surroundings.

Razin hemmed and hawed. "I am not sure," he finally said. "Perhaps it is in Moscow."

I looked up from my notes. To me, that was a big red flag.

Lincoln had already told me that every field geologist knows the location of his or her notebook at all times. It is a prized possession that is carried every day while on location. The geologist records a detailed account of every rock, grain, or clay sample collected and the exact conditions under which it was found. One could never afford to misplace it or leave it behind. The fact that Razin did not know the precise location of his field notebook disturbed me.

I tried another tack. "Can you explain to me the conditions under which you found the sample?"

"It is described in the paper," was the frosty response.

I persisted. "I would like more details about the precise geological circumstances."

Razin hemmed and hawed again, until . . . "I don't remember."

I looked up again from my notes. The big red flag had just exploded in flames.

Razin claimed that he had personally found the sample. The same sample in which he claimed to have discovered unique new minerals. Unique new materials that he had established as a holotype in the St. Petersburg Mining Museum, and that he had submitted to the International Mineralogical Association for acceptance as a new mineral.

Now he was telling me that he had no particular memory about where he found it?

I continued down my list of questions. "Do you have any more samples?" I asked.

"Maybe," he responded. "Maybe in Moscow."

Within seconds, I had a travel website pulled up on my computer screen and was checking the cost of a round-trip ticket from Tel Aviv to Moscow. Less than $500. *Not bad*, I thought.

"Would you be willing to fly to Moscow," I asked, "to look for your geological field notebook and any additional samples? I would cover the cost of the flight and the accommodations."

"Maybe," was the muted response.

The translator and I tried to figure out what "maybe" meant.

Was there a health issue? No. Was there a political issue? No. Was he reluctant for other reasons to visit Russia? No. He had traveled back and forth several times since emigrating to Israel.

It finally dawned on us that Razin might be looking for a reward.

I tried to explain to him that we were academic scientists studying minerals with virtually no market value. We were in search of tiny samples of aluminum-copper-iron alloys, samples that would be financially worthless but scientifically priceless.

We had very limited funds, I continued. We could cover his travel expenses to Moscow, but could not afford to pay a financial reward.

I was hoping Razin would appreciate the opportunity to make a contribution to science. Instead he became quiet and unresponsive. The phone call ended soon thereafter.

For the next few days, I carefully weighed all of my options and considered how I might best appeal to Razin. I asked Dov Levine, my former student, for advice. It had been twenty-five years since Dov and I had invented the concept of quasicrystals. He was now a professor at the Technion in Haifa, and a trusted colleague.

Dov put me in contact with one of his Russian friends in Haifa, who agreed to speak with Razin about his demands. *Perhaps there was a way*, I thought, *to come up with a small remuneration.*

But the request that came back through an intermediary from Razin seemed outrageous. What he wanted was far beyond what I could possibly afford to pay. Dov's friend tried to convince me otherwise. He told me that many Russian émigrés in Israel were having financial problems. Razin was a credible scientist, he said, and deserved my generosity.

But I was worried about Razin. He had managed to make a very bad impression on me during our telephone call. I had no doubt that Razin could dig up some sort of geological notebook if I sent him to Moscow. Based on our conversation, though, I was not confident it would be authentic.

I struggled for several days with the painful decision, but finally decided to break contact with Razin.

With our last-ditch effort in total ruin, Luca and I were truly despondent. How could we ever hope to learn the origin of the St. Petersburg holotype? And without finding that information, how could we ever manage to establish the authenticity of the Florence sample? And without authenticating our sample of khatyrkite, how could we ever prove that the quasicrystal we had discovered was not a fake?

I thought Luca and I had landed at the lowest of all possible low points. Unfortunately, I was mistaken. Things were about to get even worse.

TWELVE

A CAPRICIOUS IF NOT OVERTLY MALICIOUS GOD

PRINCETON AND FLORENCE, LATE APRIL 2009: Months of detective work had failed and the blue team was becoming increasingly desperate.

With nowhere else to turn, Luca and I were forced to circle back to the beginning. We still had fifty to one hundred minuscule grains left to study and each of them would be time-consuming and excruciatingly difficult to analyze. Not only were the tiny grains hard to manipulate, but many of them contained combinations of minerals that were so complicated that they would require several days, in some cases several weeks, to fully investigate.

The total amount of material was smaller than the period at the end of this sentence. But no matter how hard we worked, we knew it would be physically impossible to get through all the grains before the *Science* magazine deadline, which was now less than two months away.

It did not help that Luca and I had to pursue the investigation while juggling other research projects, regular teaching commitments, and travel to conferences and speaking engagements. For example, as director of the Natural History Museum at the University of Florence, Luca was drawn into planning an elaborate posthumous tribute for Curzio Cipriani, the former museum director. Cipriani was Luca's close friend and collaborator. He had spent half a century curating the minerals in the Florence collection.

Cipriani's widow, Marta, was also helping plan her husband's memorial. One day, after an organizational meeting, she and Luca struck up a conversation. Luca told her about the history of our project and bemoaned our current situation. We had run out of leads to pursue, and were running out of material to study. As a result, we were now in serious jeopardy of running out of time.

Marta quietly nodded and listened sympathetically. Luca mentioned that our investigation centered around a sample from the museum's Koekkoek collection. And at that, her eyes lit up. She knew that her late husband was personally responsible for that acquisition and had especially loved the Koekkoek minerals. So without hesitation, she decided to reveal one of his biggest secrets. Her husband often brought minerals home from work, Marta told Luca, in order to study them more thoroughly in a private laboratory he had assembled in their basement.

Removing samples from the museum was strictly forbidden. Even a well-respected museum director like Cipriani was not exempt from those rules. So Luca was stunned by the revelation. But he was also intrigued. If Cipriani had loved the Koekkoek collection so much, there might be important clues to be found in his personal lab. So Luca eagerly accepted Marta's invitation to visit their home the very next day.

Luca found that Cipriani, ever the professional, had been meticulous in his lab work. He had carefully logged all the details in his notebook so it was easy for Luca to quickly thumb through the well-organized pages. On one of the pages, khatyrkite was faithfully marked with the familiar number 4061, the same number found on the box of khatyrkite that Luca had originally recovered from the museum's storage room.

Looking around his mentor's secret lab, Luca was surprised to see that Cipriani had managed to gather a large collection of materials, more than a hundred samples, and had stored each of them in its own plastic box. Digging through the vast assortment, Luca uncovered a box containing a small vial labeled 4061-Khatyrkite. Inside the vial was a tiny bit of powdery material. Cipriani had apparently scraped off a few bits of the original sample and taken it home to his secret lab, where it had presumably sat undisturbed for years.

When Luca emailed me the news, I was astonished by our good luck.

Suddenly there was more khatyrkite to study! From exactly the same source that contained a natural quasicrystal!

I was certain that Cipriani's secret collection would help establish the authenticity of the khatyrkite sample and the natural quasicrystal nested within. This reversal of fortune felt like a miracle. If we were lucky, we would discover an abundance of direct contacts between the quasicrystal and the other natural minerals, which was exactly the kind of proof Lincoln and Glenn had been repeatedly demanding.

Luca and I were so convinced about the importance of the find that we decided to send it to Glenn MacPherson right away. We wanted to give the red team's biggest skeptic first crack at the pristine material, wagering it was the best way to convince him that the Florence sample was natural.

Luca mailed the powdery material to the Smithsonian the very next day. We sat back eagerly awaiting Glenn's reaction, along with the treasure trove of research he was sure to provide. *The blue team was about to prevail!* The only thing that could match my profound sense of excitement was my profound sense of relief that we were now on the verge of success.

———————

WASHINGTON, D.C., MAY 12, 2009: Ten days later, I received Glenn's email. But it was not the congratulatory note we had been expecting. I froze when I read the first line:

> I now am coming to believe in a capricious, if not overtly malicious, God.

What? No! I thought. I began to read the rest of the note. It was not good news.

> I have looked at both grains. Both particles are pieces of the Allende meteorite . . . the whole thing is enclosed in fine-grained material that can only have come from the Allende meteorite matrix or a virtually identical CV3 carbonaceous chondrite meteorite. I have spent 30 years looking at Allende, and this is it or its twin. This cannot have anything to do with the khatyrkite-cupalite material. . . . And assuming this is Allende, the purported find locality (Siberia) is something like 8000–10000 miles away from where the meteorite fell (northern Mexico). If this were a clast from the meteorite, then I'd say the presence of aluminum-copper alloys suggests that you have a piece of 6-billion-year-old alien spacecraft that became trapped in our infant solar system when it formed. . . . Paul, I do not know what to say. Based on what I now know, I'd say withdraw the paper until such time as we can come up with relevant evidence. . . . For my part, I am so flabbergasted that I am going home and have a good stiff drink. Under any other circumstances (and 40 years ago), I'd start looking around for Allen Funt and Candid Camera. Someone, somewhere is messing with my mind.

I knew the last-minute miracle from Cipriani's secret laboratory was now an unmitigated disaster. And I was right.

As one of the world's leading experts on the Allende meteorite, Glenn had immediately identified its unmistakable signature in the powdery material from Cipriani's vial. The Allende meteorite, named for its landing site near Allende, Mexico, entered the atmosphere and struck Earth on February 8, 1969. Glenn had spent years exploring every nook and cranny of Allende, because of the secrets it carries about the birth of our solar system.

Some cosmologists believe the universe sprang from nothing in a big bang that occurred 13.8 billion years ago. Others think that the bang might actually have been a bounce, a transition from an earlier epoch of contraction to the current period of expansion, in which case the universe could be much older. In either case, cosmologists agree that 13.8 billion years ago, the universe was much hotter and denser than the core of the sun. Space was filled with a hot gas teeming with freely moving protons, neutrons, and electrons. As space expanded and the hot gas cooled, the elementary particles that had been hurtling around freely began to clump together to form atoms and molecules, dust, planets, stars, galaxies, clusters of galaxies, and clusters and clusters of galaxies. About nine billion years later, in the galaxy known as the Milky Way, our solar system began to form from a dust cloud composed of the remains of earlier generations of stars. Enough matter fell into the center of the dust cloud to form our emergent sun. The remaining dust swirling around the sun slowly coalesced, condensed into the planets, asteroids, and other objects we observe orbiting the sun today.

The Allende meteorite, along with other meteorites known as "CV3 carbonaceous chondrites," formed more than 4.5 billion years ago at the birth of our solar system, just as the sun was igniting. Samples are highly coveted because they provide scientists with extremely valuable information about the chemical and physical conditions that existed at that time.

Glenn had studied Allende samples long enough and thoroughly enough to recognize one in his sleep. So he was understandably shocked to find that Cipriani had mistakenly slipped powdery material from the well-known Allende into a vial of a very unremarkable material labeled 4061-Khatyrkite. How could Cipriani have possibly mixed up the Florence museum sample with such a famously recognizable material? Whatever the circumstances, Glenn found it inexcusable.

Luca was dumbfounded and mortified. I was more sanguine. To me, it was merely the latest downturn in an already wildly fluctuating investigation. After all, we could never know what experiments Cipriani had been conducting in his home laboratory. Yes, the 4061 label was a match to the museum label. But all of our meaningful data had been drawn from the *other* 4061 sample, the carefully preserved museum sample of khatyrkite that was clearly not a piece of the Allende meteorite.

Glenn viewed the situation differently. He made it clear to us that he considered the fiasco a shameful point of no return. He had now lost confidence in *everything* originating in Florence. Why should he trust the original museum sample? The entire museum, he asserted, could be riddled with misidentified samples and fakes.

Worst of all, Glenn decided that he was now vehemently opposed to publishing the *Science* paper and lobbied to make sure that the red team presented a unified front. He sent Lincoln a copy of his incendiary email about a capricious if not malicious God, effectively suggesting that he join the resistance.

PRINCETON, MAY 15, 2009: I knew that the *Science* manuscript was already in the hands of the copy editor and speeding toward publication. So Glenn's reaction to the Cipriani incident presented a dilemma. As team leader, I would have to make a difficult decision that would affect everyone's professional reputation: Publish or withdraw the *Science* manuscript?

As skeptics, the red team's mission was to help prevent us from fooling ourselves, as Feynman had warned. Lincoln and Glenn were terrific collaborators. But their friendly opposition was mushrooming into open revolt.

How could we possibly publish our discovery without their support? I thought.

Since neither Lincoln nor Glenn were listed as coauthors, it was not their decision to make. At the same time, Luca and I had no desire to discount their opinions. They were knowledgeable, important advisors who had contributed greatly to every aspect of the research we had conducted since the original discovery.

There were, however, two other official coauthors to consult, Peter Lu and Nan Yao. Ten years ago, Peter had helped me canvass the worldwide database of minerals. Five months earlier, I had collaborated with Nan Yao, the brilliant microscopist and director of the Princeton Imaging Center, in making the initial discovery.

Peter advocated that we publish immediately. He expressed confidence that the sample was natural based on his impressions of the only available images of the original piece of museum khatyrkite. Nan declined to vote, and deferred to the blue team's scientific judgment.

I weighed all the opinions against the evidence in hand and quickly settled on a course of action. But I purposely waited before reaching out to Lincoln and Glenn. Tempers were running hot, and I wanted to give the red team a chance to cool down so we could hold a dispassionate discussion.

In a series of meetings and phone calls, I reminded them about the thick notebook of observations and data that we had collected on the Florence museum sample. All of it favored the conclusion that the sample was natural.

Both Lincoln and Glenn conceded that was true.

Next, I argued, the Cipriani incident was irrelevant. Perhaps it was true that Cipriani brought samples home from the museum and subsequently bungled their handling. But ultimately, the Cipriani experience did not prove or disprove anything. We should just ignore it, I contended. All of the observations and data in the *Science* paper were strictly limited to the original sample of khatyrkite that had been carefully tracked and properly curated in the Florence museum.

Of course, Glenn had become suspicious of everything originating in Florence, including the museum sample. When pressed, though, he had to admit that there was not a shred of evidence that anything in the museum was tainted.

Finally, I summarized what appeared to me to be the essential point on which everyone could agree: The preponderance of the evidence indicated that the Florence sample of khatyrkite and the quasicrystals within it were natural. Neither Lincoln nor Glenn could disagree with that conclusion.

But this was the full extent of what we were claiming in the paper, I argued. We did not claim absolute proof that the quasicrystal was natural. In accordance with their concerns, we included a warning that the presence of minerals with metallic aluminum, including the quasicrystal, was a serious challenge to explain. We claimed no definitive explanation, we were merely presenting all of the evidence to date. We acknowledged that the presence of aluminum could mean the sample was a by-product of human activity. At the same time, though, we also presented substantial experimental evidence favoring the alternate theory, which was the admittedly daring hypothesis that the quasicrystals had formed naturally.

I have always believed that a paper that honestly presents the supporting evidence along with a clear warning about its limitations is scientifically responsible. I also believed that publishing our discovery would allow other scientists to weigh in with more evidence or with better ideas that could help explain the baffling existence of the Florence sample.

Luca completely agreed with my analysis. Lincoln and Glenn, however, strongly disagreed with me and their stubborn opposition revealed the true nature of the scientific conflict. They were concerned that a theoretical physicist, meaning me, was applying a lower standard than the petrologist and the meteorite expert, meaning themselves. In their view, the paper should not be published unless and until we could definitively rule out the possibility that the metallic aluminum alloys were man-made. No matter how long it would take to accomplish that.

For Lincoln and Glenn, the remaining uncertainty outweighed the preponderance of evidence. I finally realized that there was no amount of careful writing or full disclosure on my part that could ever compensate. With so many unanswered questions, they both expressed the concern that the paper could turn into a professional embarrassment and, at one point, explicitly asked that I withdraw their names from the acknowledgments or at least change the acknowledgments so that it was clear that they disapproved of the conclusions.

But by this time, the manuscript had been sent to the printers. The journal declined to perform a partial edit and refused to remove the acknowledgments. The only alternative the editors offered was to scrub the article altogether.

So I had arrived at the tipping point. Pulling the article at the last moment would be a consequential decision. Withdrawing from publication is usually interpreted as a sign of trouble, and bad news has a way of spreading quickly within the scientific community. I knew stopping publication would raise suspicions and threaten the future credibility of the project.

As I wrestled with the final decision, I revisited the events of the last five months. More than any scientific venture I had ever been involved with, the natural quasicrystal investigation that Luca and I had pursued had been an unpredictable affair with an inordinate amount of gut-wrenching twists and turns. I could certainly appreciate that Lincoln and Glenn were accustomed to more disciplined investigations.

I could also understand why the red team was suspicious of fakery. In their view, metallic aluminum was simply impossible in the natural world. But in my opinion, the endless hours of laboratory work in Florence, D.C., and Princeton had substantially eliminated every idea about how the sample could have been artificially produced. A mineral of natural origin, even if we could not explain all the details surrounding it, stood as the most plausible explanation.

Every time I went over the arguments, I came to the same conclusion: We should proceed with the paper. It was responsibly written, and the cost of withdrawing it from publication had become much too high. There was no doubt in my mind that I was making the right decision. I firmly overruled the objections of the red team.

Our paper, titled "Natural Quasicrystals," was published in *Science* magazine on June 5, 2009. It was coauthored by Luca Bindi, Nan Yao, Peter Lu, and me with acknowledgments to Lincoln Hollister and Glenn MacPherson.

The aftermath was bittersweet. On the one hand, the *Science* magazine article received a substantial amount of international attention. There was no sign of any skepticism, which was surprising given the controversial nature of what we were reporting. Luca and I were ebullient. At the same time, I could not help noticing that the red team had suddenly gone silent. Lincoln and Glenn gave us no indication that they were aware of the publication and chose not to comment in any of the many follow-up articles that were published in multiple outlets all over the world.

REPORTS

Natural Quasicrystals

Luca Bindi,[1] Paul J. Steinhardt,[2*] Nan Yao,[3] Peter J. Lu[4]

Quasicrystals are solids whose atomic arrangements have symmetries that are forbidden for periodic crystals, including configurations with fivefold symmetry. All examples identified to date have been synthesized in the laboratory under controlled conditions. Here we present evidence of a naturally occurring icosahedral quasicrystal that includes six distinct fivefold symmetry axes. The mineral, an alloy of aluminum, copper, and iron, occurs as micrometer-sized grains associated with crystalline khatyrkite and cupalite in samples reported to have come from the Koryak Mountains in Russia. The results suggest that quasicrystals can form and remain stable under geologic conditions, although there remain open questions as to how this mineral formed naturally.

Solids, including naturally forming minerals, are classified according to the order and rotational symmetry of their atomic arrangements. Glasses and amorphous solids

number have icosahedral symmetry, but other crystallographically forbidden symmetries have been observed as well (1, 4). Among the most carefully studied is the icosahedral phase of

17. D. Levine, T. C. Lubensky, S. Ostlund, S. Ramaswamy, P. J. Steinhardt, J. Toner, *Phys. Rev. Lett.* **54**, 1520 (1985).
18. B. Dam, A. Janner, J. D. H. Donnay, *Phys. Rev. Lett.* **55** 2301 (1985).
19. E. Makovicky, B. G. Hyde, *Struct. Bonding* **46**, 101 (1981)
20. We are indebted to L. Hollister and G. MacPherson for their critical examination of the results, especially regarding the issue of natural origin. We also thank P. Bonazzi, K. Deffeyes, S. Menchetti, and P. Spry for useful discussions and S. Bambi at the Museo di Storia Naturale for the photograph of the original sample in

FLORENCE, JULY 3, 2009: About a month after the publication of our discovery of the first-ever natural quasicrystal, and two years after we started working together, Luca and I finally had the chance to meet in person. I was able to visit him in Florence during a trip to deliver a series of lectures in Europe on a different topic.

We greeted each other with a warm hug. After hundreds of daily Skype chats about everything from scientific problems to family issues, meeting Luca felt like a reunion with an old friend. It was clear that the intensity of our working relationship had forged a strong bond between us. Luca was taller and even more athletic than I had imagined. The bubbly enthusiasm and sincere warmth he managed to convey over the Internet were all the more evident in person. The two of us are from different cultures, different generations, and different scientific backgrounds. But we had discovered ourselves to be truly kindred spirits.

Luca took me on a brief tour of the university museum and proudly showed me the beautiful new mineral displays he had designed. We settled down to work in his office, and spoke nonstop for several hours about the state of the investigation and what should be done next.

Although the *Science* article was now successfully published, we were both keenly aware that we had fallen short of the high expectations of the red team. It made little difference that the scientific journals, the general press, and the readership had accepted our result. We would not be satisfied until one of two things happened: We either had to convince Lincoln and Glenn that we were right, or they had to convince us we were wrong. So that meant our investigation would continue. Luca and I agreed we would sift through more grains, look for more clues, conduct more experiments on the grains we had already studied, read more about aluminum alloys, and explore more theories about how the Florence sample may have formed.

I admitted to Luca that I had anticipated that the *Science* article would help prime the pump. I had hoped, for example, that it might motivate geologists from around the world to check their own mineral collections for natural quasicrystals or, better yet, send us their samples to study. But to my disappointment, no one reached out to us. We would need to do the work ourselves.

Our intense discussion continued through lunch, until we reluctantly had to go our separate ways. I affectionately hugged Luca goodbye and left Florence with an even greater appreciation and admiration for my Italian colleague.

I had begun to sense a growing coldness from Lincoln and Glenn in our emails, which had now become sporadic. And as it turned out, my instincts were correct. A few weeks after I returned home from Europe, Lincoln expressed his frustration to me in a stern note:

> I believe that the sample you have been working with is not natural.
> I feel I am up against a wall of diminishing returns to determine its origin.

Lincoln explained that he did not want to continue working with us unless we could somehow find a completely fresh sample from some other source. Glenn's withdrawal from the project was implicit.

I felt sad and discouraged as I read Lincoln's email. It was certainly true that the red and blue teams had experienced their fair share of disagreements. But like all good scientific disagreements, our debates had always remained civil and had never devolved into personal attacks. I always believed that Lincoln and Glenn were crucial to our investigation so I was determined to keep them involved by one means or another. The fact that we were currently weathering a difference of opinion did not alter my respect for them one iota.

But where could the next breakthrough come from? And in the absence of any imminent news, what could I possibly do to bring Lincoln and Glenn back on board?

THIRTEEN

THE *SECRET* SECRET DIARY

FLORENCE, SEPTEMBER 2009: It had been nearly three months since our *Science* article had announced our discovery. A full summer's worth of investigation had passed and no one on our team had anything remotely interesting to report. Luca, Nan Yao, and I had spent long hours toiling away in the lab but were not making any progress.

Then, just when it seemed that our project might grind to a permanent halt, the most unexpected thing happened. It did not surface in the lab. Or at a conference. Or through an exchange with other scientists. The catalyst was wine and pasta.

Luca was enjoying dinner in Florence with his sister Monica and her friend Roberto and was regaling them with the most dramatic highlights of our story. By now, it was a long tale: The invaluable sample of khatyrkite Luca found tucked away in his museum's storage room, the unexpected discovery of a natural quasicrystal in the Princeton lab with Nan Yao, the embarrassingly fake samples we discovered in private collections, the untouchable holotype locked away in a St. Petersburg museum, the untrustworthy Russian scientist we tracked down in Israel, the inexplicable mix-up with the famous Allende meteorite, along with endless rounds of inconclusive testing and debate.

Luca explained that we had traced the original khatyrkite sample

1

Quasicrystal tilings can be made with any symmetry—for example, this pattern, comprised of five different tile shapes, has eleven-fold symmetry.

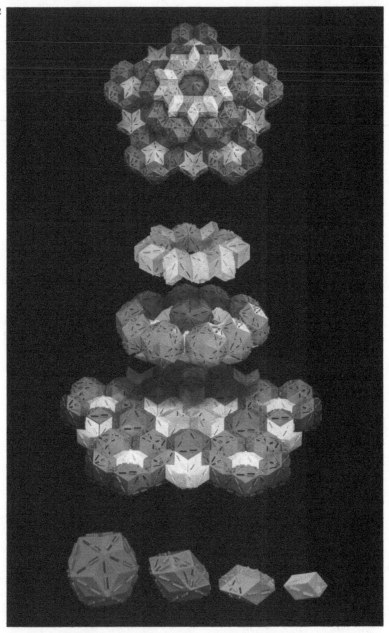

A three dimensional model (top) of an icosahedral quasicrystal built from layers (middle) composed of four different building blocks (bottom). The flanges and slots on the blocks force them to assemble in a quasicrystal arrangement.

The remarkable tiling on the Darb-i Imam shrine in Isfahan, Iran, (top) can be viewed as a quasicrystal tiling (bottom left) composed of three shapes known as girih tiles (bottom right).

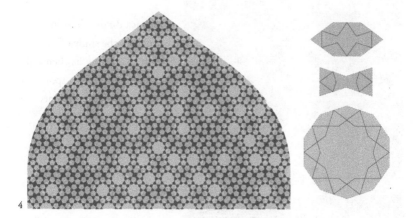

The khatyrkite sample from the Florence museum in its original box (held by putty), next to a 5-cent euro coin for scale. Below, the sample is enlarged about ten times.

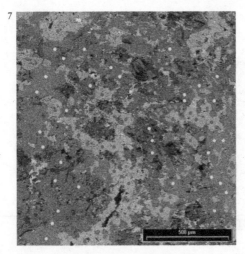

Luca Bindi's electron microprobe study of the sample, measuring its chemical composition from spot to spot. Yellow represents khatyrkite, $CuAl_2$; red is cupalite, CuAl; green is a crystal mix of AlCuFe; blue is the first natural quasicrystal, icosahedrite, $Al_{63}Cu_{24}Fe_{13}$.

.8

Top: The two behemoths and the expedition team (left to right): Bogdan Ma-kovskii, Glenn MacPherson, Will Steinhardt, Chris Andronicos, Marina Yudov-skaya, Luca Bindi, Viktor Komelkov, Olya Komelkova, Paul Steinhardt, Sasha Kostin, Valery Kryachko, Michael Eddy, Vadim Distler, and, in the foreground, Bucks. *Bottom:* A behemoth crash; Bucks, in close-up; and a Kamchatka bear.

9

10

11

12

13

14

Top: Russian scientist Valery Kryachko in Kamchatka (with Paul Steinhardt over his left shoulder) reviews the expedition route with the team. *Middle:* Will Steinhardt along the excavation site at the Listvenitovyi Stream. *Bottom:* Luca Bindi and Paul celebrate their 2011 arrival at the spot along the stream where Valery discovered the Florence sample in 1979.

Top: Valery Kryachko pans for samples; Glenn MacPherson examines a rock for evidence of a meteor strike; Valery and Luca Bindi check grains; Chris Andronicos and Paul Steinhardt (wearing mosquito gear) map geological features. *Bottom:* The view from the campsite at midnight.

Top: Marina Yudovskaya and Mike Eddy during a mapping expedition; Will Steinhardt preparing to dig at the stream; Chris Andronicos armed with a rifle to defend against bears. *Bottom:* The team celebrating its final night in the field, with Paul Steinhardt holding a flare aloft.

in the Florence museum back to a Dutch mineral collector. But unfortunately for us, the trail ran cold in Amsterdam. Luca's dinner companion, Roberto, lived in Amsterdam and was intrigued by that detail. He nodded when he heard the collector's last name was Koekkoek. It was not so surprising that he was hard to locate, according to Roberto, because that was a relatively common last name. In fact, Roberto had a neighbor named Koekkoek. The elderly woman lived just down the street, and he often helped carry her packages home from the grocery store. He promised Luca he would ask for her advice.

By now, Luca and I had already spent months scouring Amsterdam. There was little chance, Luca figured, that a random acquaintance of Roberto's, someone with no connection to our story, could be helpful. But he was wrong.

Within twenty-four hours, Roberto had returned to Amsterdam and dashed off an email to Luca. Not only did his neighbor know of Nico Koekkoek, the two of them had been *very* well acquainted. She was, in fact, his widow!

The unexpected news hit us like a thunderbolt. Luca immediately bought a ticket for the two-hour flight to Amsterdam and shot me a quick email to say he was going to try to interview the elderly woman as soon as possible.

"I feel myself as a CIA agent," Luca wrote.

AMSTERDAM, SEPTEMBER 2009: Luca swooped into Amsterdam the next day with high expectations. He excitedly headed for Roberto's neighborhood and the Koekkoek apartment, where, to his surprise, he and Roberto were abruptly stopped in their tracks by a stubborn brick wall named Debora Koekkoek. The eighty-year-old was apparently taken aback by the unsolicited visit and, to Luca's dismay, categorically refused to cooperate. She was unwilling to share any of her family's private information with an unknown Italian, no matter how charming or how persuasive he may be.

To his credit, Roberto tried his best to salvage the situation. The only choice, he determined, was for Luca to disappear from sight so that he could try to speak privately with his neighbor. Luca reluctantly agreed, and waited sullenly at a nearby café.

How could Roberto be expected to discover anything useful, Luca wondered. *Two days earlier, he had never even heard about our quest.*

As Luca steamed, the discussion between Roberto and Debora turned into a battle of wills. Whenever Roberto asked Debora about her husband's collection, she insisted that she knew virtually nothing about it. She was willing to admit that her late husband had traded in minerals and seashells. She also knew that he had liquidated everything in his mineral inventory in 1990 in order to focus exclusively on collecting seashells, which he found more appealing. That was it. That was all she knew. End of story. No matter how many different ways Roberto tried to ask about the mineral collection, Debora pleaded ignorance.

Finally, perhaps because Roberto said something that jogged her memory or perhaps to put an end his persistent questions, Debora shyly volunteered a crucial piece of new information. Even though her husband had sold off his mineral collection, she told Roberto, he had never disposed of the secret diary in which he kept a record of his purchases. And she still had it.

After some gentle persuasion, Debora agreed to let Roberto peek at the secret diary. Sure enough, he quickly found an entry for khatyrkite, which Nico Koekkoek simply described as "ore from Russia." He had also dutifully recorded that he had obtained the sample on a trip to Romania.

The entry further explained that Koekkoek purchased the sample in Romania in 1987 from someone named Tim. There was no last name or contact information.

A mineral dealer named Tim? In Romania? Tim the Romanian?

Roberto jotted down some notes, said his goodbyes to Debora, and conveyed the news to Luca, who passed it on to me. Luca and I surmised that Tim was most likely a mineral smuggler. He and Koekkoek were apparently doing business together in the late 1980s, a time when Romania was still in the grip of a communist dictator and considered a Soviet satellite. Sneaking natural minerals out from behind the Iron Curtain at that time would probably have been considered a serious offense.

PRINCETON AND ROMANIA, OCTOBER 2009: The next step was sure to be simple, I thought. Compared to finding Leonid Razin in Israel or the widow of a Dutch mineral dealer in Amsterdam, tracking down Tim the Romanian was going to be a snap.

After all, I thought, *how many smugglers could there be in Romania named Tim?*

My optimism was ill-founded. Although we sent out an all-points bulletin to contacts in Romania and collectors worldwide, it seemed that no one had ever heard of Tim the Romanian.

As we continued to widen our search for Tim, there was a glimmer of hope on a different front.

One of the problems we had been grappling with from the onset of the investigation was that we had only two tiny specks of grains to study and an unusually limited amount of information about the rock they came from. There was one highly magnified image of a single slice of the original sample, which showed a complex configuration of aluminum-copper alloys and silicate minerals. But after taking that image, Luca had pulverized the slices in order to try to extract the specks that he sent to Princeton for me to examine. And of course, those were the specks that turned out to include the first-ever natural quasicrystal.

Lincoln Hollister was constantly complaining that there was only one image to study. Whenever we would meet, he would repeatedly emphasize that Luca had made a big mistake by pulverizing the Florence sample, especially without first having taken a more extensive series of images at different magnifications with his electron microscope. The images could have shown that the quasicrystal and other aluminum-copper alloys were enmeshed with, or had multiple contacts with, silicate minerals known to be natural. By identifying the contacts and perhaps finding examples where the metal and silicates reacted chemically with one another, we would have strong evidence that the quasicrystals were natural, too. Unfortunately, after everything was pulverized, the grains were too small to provide convincing evidence one way or the other.

The criticism was particularly difficult for me to accept because I was painfully aware that it was undeserved. The truth was that Luca *had* taken a full series of electron microscope images. The problem was that they had been lost. After Luca took the appropriate series of images, his Florence laboratory was hit with a crippling one-two punch—the electron microscope broke and the hard drive crashed beyond repair. Luca's lab proceeded to replace the microscope and the hard drive. The remains of the broken equipment were unceremoniously shoved into a corner and abandoned. Luca's photos were lost on the mangled hard drive.

Luca had immediately told me about the disastrous events and was understandably upset. His greatest fear was that the problems would give Lincoln and Glenn the impression that his laboratory was substandard and amateurish. Luca thought explaining the truth, that he was the victim of a random mechanical breakdown, would sound pathetic, along the lines of "the dog ate my homework." So he swore me to silence, having decided it would be better for him to suffer a barrage of criticism from the red team rather than to offer a lame-sounding excuse.

I respected Luca's decision, but suggested we make a quiet effort to see if a data recovery specialist could reclaim any of the lost images. Unfortunately, the expert we consulted did not offer much hope. He thought there was little chance of success because significant portions of the hard drive had been irreparably damaged by the crash. Disheartened once again, the two of us turned our attention back to the main investigation and forgot all about the salvage effort.

Months later, just as the search for Tim the Romanian mineral smuggler was reaching its depressing conclusion, Luca received an unexpected message from the computer wizards. They had somehow managed to recover a handful of images from the crippled hard drive.

The good news about the recovered images was a convenient icebreaker for me to use to try to reestablish rapport with Lincoln and Glenn. They were surprised to hear the full truth about Luca's secret equipment failure. And they were fascinated by the set of recovered images, one of which is shown below.

Lincoln and Glenn were probably expecting clear evidence that

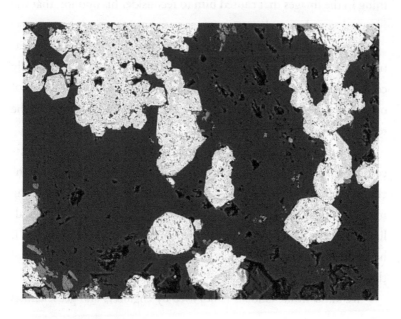

the aluminum-copper alloys were artificial, just as they always suspected. But instead, as Glenn noted in an email, the images were a bit of a surprise:

> There is vastly more complexity here than seen in any of previous photos. The term 'dog's breakfast' applies!

Glenn has a penchant for picturesque language. And from that point on, "dog's breakfast" was adopted as part of our team's internal jargon. The phrase itself is British slang for a meal so badly ruined that it would only appeal to the four-legged member of the family. That, of course, implies quite a mess indeed, given that traditional British fare includes such things as blood pudding and jellied eels.

In this instance, Glenn was trying to convey his impression that the images were a jumble that was hard to interpret. They were unlike anything he had ever studied before. But he did not point out anything in the images that caused him to reconsider his opinion that the Florence sample was slag.

Luca and I, on the other hand, detected several significant features that Glenn appeared to have missed. First of all, slag typically contains certain telltale features, such as bubbles or bits and pieces of other common industrial materials. Yet none of this was visible in the recovered images.

Secondly, the interfaces between the metal seen in the light-colored materials, and the silicate seen in the darker materials, included some straight edges. The surrounding silicate, containing mixtures of silicon, oxygen, and other components, was also crystalline. This configuration between two minerals could only occur if both materials were first entirely melted into a liquid mixture and then slowly cooled.

We knew from standard tables used by engineers and geoscientists that the cooling silicates would have been the first to crystallize, at about 1500° C. The copper-aluminum alloys would have crystallized later, at about 1000° C. That provided us with quantitative information about the high temperatures the dog's breakfast had been subjected to.

The fact that the metal had melted but not reacted with oxygen in the melted silicate was also significant. Melted metal is normally highly reactive because the metal atoms are free to move around and chemically react with any oxygen atoms that happen to be in their surroundings. But here was a clear-cut case of melted aluminum making contact with an oxygen-rich liquid silicate and not reacting.

A logical explanation was that the metal solidified ultra-rapidly, before it could react with the oxygen bonded to the silicate. Ultra-rapid cooling would also explain the oddly contorted shapes. Yet rapid cooling at such a high rate would not normally occur as part of any natural process on the surface of the Earth or in an ordinary laboratory.

Thanks to the images recovered from the hard drive, Luca and I were able to conclude that the list of possible sources of the Florence sample was narrowing. Everything now pointed to a natural origin.

AMSTERDAM, NOVEMBER 2009: Buoyed by our recent success recovering the computer images from the broken hard drive but frustrated by our failure to find Tim the Romanian, Luca and I decided to try something that had almost no chance of succeeding. Luca would go back to Amsterdam and ask Debora, the recalcitrant widow of the Dutch mineral collector, if she knew anything about Tim.

This time around, Debora was more comfortable with Luca and invited him into her home. After exchanging pleasantries, he got right to the point: Had she ever heard her husband Nico refer to anyone from Romania named Tim?

Debora's answer was straightforward and direct: No.

Luca persisted, though, remembering how long it had taken Roberto to extract information from her. He explained to Debora that her late husband had mentioned Tim by first name in the diary she had shown Roberto. So he was probably someone very familiar.

Perhaps she recalled a story her husband told her about his adventures in Romania? No, Debora insisted.

And so the conversation went. No matter how many different ways Luca tried to broach the topic, Debora's answer was always the same. She had never heard her husband mention anything about anyone named Tim.

Then, just as Luca was about to give up, Debora quietly made a startling confession. Her husband kept a second diary, a *SECRET* secret diary. The second diary was apparently Koekkoek's way of keeping track of purchases he knew had questionable legality, including minerals obtained under suspicious circumstances. He did not want to create a paper trail about those transactions in his official records. Debora had apparently been too nervous or too embarrassed to have mentioned it earlier. But now, she quickly retrieved the *SECRET* secret diary from the next room and gave it to Luca. Once the diary was in his hands, it did not take long for him to find what he was looking for.

The entry translated as follows:

Most metals minerals present in my small collection of fragments were conveyed to me through Tim.

The opening line was consistent with Koekkoek's first diary. But the *rest* of the entry offered something new:

Tim is obtaining minerals in the laboratory Rudashevsky. The minerals were given primarily to Tim by L. Razin (director at major center in Russia) a man who worked in the laboratory Rudashevsky.

> tot het einde. De meeste metalen mineralen
> aanwezig in mijn kleine verzameling van fragmenten
> werd aan mij door de Jevon heen door Tim. Tim verkrijgen
> van mineralen van het laboratorium Rudashevsky De
> mineralen werden gegeven in de eerste plaats aan Tim L.
> Razin (directeur van een belangrijk centrum in Russia) of
> uit een man die werkte in het laboratorium Rudashevsky.

It must have been hard for Luca to maintain his composure. All of the names in the *SECRET* secret diary were well known to us by now. Two of them were scientists who featured prominently in the khatyrkite story. Evidence of their connection to Nico Koekkoek came as an utter shock.

Razin. Based on Koekkoek's diary entry, the Florence sample was not just chemically similar to the St. Petersburg holotype. It came directly from the same source, Leonid Razin, the uncooperative scientist we tracked down in Israel who claimed to have personally discovered the sample.

Rudashevsky. Luca and I knew Nikolai Rudashevsky to be the electron microscopist who had worked with many Russian mineralogists in his St. Petersburg lab. In exchange for his efforts, Rudashevsky was often given credit as coauthor on their scientific papers. In 1985, one of the papers he coauthored with Razin described the discovery of khatyrkite and cupalite.

Koekkoek's mention of the two Russian scientists meant that both the Florence sample and the holotype sample in the St. Petersburg Mining Museum must have come from the same place—a remote area of Eastern Russia that was far from any aluminum foundries or sophisticated laboratories that could have produced the samples artificially.

The St. Petersburg holotype, of course, was already certified by the International Mineralogical Association to be of natural origin. If the two samples came from the same place, it was reasonable to suppose that the Florence sample was natural, as well. Luca and I were certain that Koekkoek's *SECRET* secret diary was a watershed moment that was about to help us prove our case for the first-ever natural quasicrystal.

FOURTEEN

VALERY KRYACHKO

PRINCETON AND FLORENCE, NOVEMBER 2009: According to Nico Koekkoek's *SECRET* secret diary, Leonid Razin was the source of the Florence khatyrkite sample. He was also the lead author of the paper describing the St. Petersburg holotype.

But how did Razin obtain the two samples?

Razin claimed that he had recovered the minerals in 1979 in a remote, uninhabited region of Far Eastern Russia. But I knew that the expedition would have entailed considerable inconvenience. Based on my experience with Razin, I doubted his claim.

At the time, Razin was head of the Soviet Union's Institute of Platinum in Moscow. Colleagues believed he had strong family connections to both the KGB and the Communist Party, which controlled every aspect of daily life in the former Soviet Union. Those who may have been personally close to operatives in the KGB, the security agency that later became the FSB, were thought to be capable of destroying other people's lives.

It seemed unlikely that such a high-ranking, politically connected person accustomed to city life would decide to embark on a taxing mission to such a desolate location. I was convinced the actual fieldwork must have been done by an underling.

But who was it?

ST. PETERSBURG, NOVEMBER 2009: Koekkoek's diary suggested that Razin had been working with Nicolai Rudashevsky's lab when a sample of khatyrkite had been smuggled out of the country. I managed to track down Rudashevsky, who was by then in his eighties and still living in St. Petersburg with his family. His son Vladimir was fluent in English and served as our intermediary. Vladimir had followed his father's footsteps into science and had become a successful entrepreneur in the minerals industry. He was proud of his father's accomplishments, understood the scientific ramifications of our research, and was fully supportive of our impossible quest.

Vladimir spent hours talking to his elderly father about our investigation, but failed to jog his memory about the work he had performed more than thirty years earlier on khatyrkite and cupalite. That was not completely surprising; Razin's samples were not particularly noteworthy at the time, and Rudashevsky's lab had conducted hundreds, if not thousands, of tests on other materials since.

Stymied, Vladimir tried to come up with another way to help us. I had told him about my surprisingly difficult encounter with Leonid Razin. Acting in his father's stead, Vladimir offered to call Razin directly to see if he could make some headway. I warily agreed, hoping that a personal entreaty from a former colleague's son who was himself a fellow scientist would coax Razin into cooperating.

But Vladimir's conversation with Razin was even more disastrous than mine. Razin had originally tried to stonewall me because I would not pay his exorbitant fee. Now, he was enraged to learn that Vladimir was trying to help me get the information for free. Vladimir told me that Razin had become so furious during their telephone call that he had tried to intimidate him by reminding him about his political connections.

I was taken aback by the threat. I had been warned by Razin's former colleagues that he could be malevolent. So now I became worried that a harmless scientific investigation might result in harm to the Rudashevsky family. Even though he was living in Israel, it seemed possible that Razin might have a long reach.

When I tried to apologize to Vladimir for exposing him and his father to potential danger, he just laughed. He told me he was confident that Razin was making an idle threat. And indeed, nothing ever happened to the Rudashevskys. Even so, Luca and I took the matter to heart and vowed to avoid any future contact with Leonid Razin.

FLORENCE AND PRINCETON, DECEMBER 2009: Luca and I must have read the original scientific paper announcing the discovery of khatyrkite and cupalite a thousand times. The first paragraph always seemed strangely cryptic.

It mentions a person (the first highlight below) named V. V. Kryachko, who discovered some unidentified grains while panning in the Listvenitovyi Stream (the second highlight). The paragraph never elaborates about the panning or explains exactly how it was related to the discovery of khatyrkite and cupalite. Secondly, the mysterious Kryachko character was never mentioned again and was not listed as a coauthor. That omission would seem to indicate that Kryachko was not significant to the discovery. So, if that was the case, why mention him or her at all? Thirdly, despite our most

Среди природных образований впервые обнаружены соединения алюминия с медью и цинком. Они находятся в тесном срастании и представлены мелкими (размером от долей до 1.5 мм) неправильной формы, угловатыми стально-серовато-желтыми металлическими частицами, внешне схожими с самородной платиной. Эти частицы встречены в черном шлихе, отмытом В. В. Крячко из зеленовато-синей глинистой массы элювия серпентинитов. Шлих отмыт непосредственно в полевых условиях, из обнажения коры выветривания серпентинитов небольшого массива ручья Лиственитового. Лабораторной обработке шлих не подвергался.

rigorous searching, we had never been able to find the mysterious Listvenitovyi Stream on a map.

Did this person or this place really exist?

A leading Russian academician told us there was a very simple possible explanation: Maybe Kryachko was a fictional character and the Listvenitovyi Stream was a fictional place?

Razin was director of the Institute of Platinum, our contact reminded us. He was searching for valuable ore metals and may have been trying to obscure all the details for competitive reasons. Even though khatyrkite and cupalite were not marketable, Razin's competitors would have guessed that they were discovered during the Institute's search for platinum. If Razin had accurately documented the events and locale, adversaries might have been able to glean enough information to raid the valuable deposits. So it would be practically mandatory for Razin to create a fictional story to throw his competitors off track.

The explanation sounded plausible in a strange sort of unscientific way.

We double-checked the "fictional" theory with another Russian academician, who promptly disagreed. He assured us that Mr. Kryachko was not an imaginary character, but rather a known mineralogist. Sadly, he reported, Kryachko had died several years earlier.

With two disparate theories now in hand, we checked with a third Russian academician and received yet a third explanation. He reported that Kryachko was a Chukchi, one of the indigenous people in Chukotka. Kryachko would have been hired to assist Razin's expedition and would have long since returned to his village in the tundra. Finding him would be a hopeless endeavor, we were told. And since he was merely a helper, he would not have any useful information for us.

So according to the experts, V. V. Kryachko was either a fictional person, a deceased person, or an untraceable person. Whatever the truth, it made no difference. The bottom line was that we were never going to make contact. Luca and I moved on.

Months after giving up on Kryachko, Luca and I accidentally stumbled across the name again in a different context. We were poring through a vast literature on Russian mining ores when Luca spotted an obscure article on platinum-group minerals discovered in the Koryak Mountains. The paper was coauthored by none other than V. V. Kryachko.

The same exact name? In the same exact region? Studying closely related minerals?

We were sure it was not a coincidence. But there was no information listed about V. V. Kryachko, not even a professional affiliation. So there was no way of telling if he was still alive.

We turned to the paper's coauthor, Vadim Distler, whom we identified as a leading researcher at IGEM, the Russian Academy of Sciences Institute of Geology of Ore Deposits, Petrography, Mineralogy, and Geochemistry in Moscow.

My email to Vadim went unanswered for several weeks, long enough for me to worry that we had reached another frustrating dead end. When he finally responded to my note, Vadim apologized for the delay and explained that he had been unable to get to his office because of a personal illness and the harsh winter weather in Moscow. We made an appointment to talk on the telephone, and I recruited a Russian colleague at Princeton to help translate.

When we finally connected, I began by trying to find out if we had the right person. Did Vadim know if his coauthor, V. V. Kryachko, was the same person mentioned in Razin's paper? I was desperate for good news, because we were running out of ideas and this was our only potential lead.

I held my breath waiting for Vadim's response. "Da!" was his reply. I threw up my arms in victory.

The rest of the conversation was a gold mine. V. V. Kryachko was Valery Kryachko. Decades earlier, Vadim had been Valery's PhD thesis advisor at the Institute of Geology of Ore Deposits. One summer before Valery had entered graduate school, Leonid Razin had offered him an opportunity to gain some valuable field experience. In 1979, Razin had sent young Valery to the Listvenitovyi Stream to search for platinum.

I felt a smile come over my face as the sentence was translated. Here was the simple explanation that connected all of the seemingly disparate elements.

Leonid Razin, the Institute of Platinum, Valery Kryachko, a student sent on an expedition, the Listvenitovyi Stream.

I dreaded the answer to the next question but had to ask: Is Valery still alive? After what felt like an interminable pause, I heard Vadim's response, "Da!"

I could barely contain my excitement. *Valery is alive?* I thought. If I could manage to find him, he might be able to explain how the samples had been found. We could finally answer all the questions Razin had been trying to bluff his way through.

While all these thoughts were racing through my mind, Vadim continued speaking with the translator. "Valery is planning to come to Moscow to visit me at the end of the month," he said. "Would you like me to have him contact you?"

I looked at the translator in disbelief. Are you kidding? I laughed. Tell him "Yes, yes, yes, a thousand times yes!"

PRINCETON AND MOSCOW, RUSSIA, JANUARY 7, 2010: One year and five days after we discovered the first natural quasicrystal in a Princeton laboratory, I managed to establish direct communication with Valery Kryachko, the scientist who had pulled the sample out of the ground more than thirty years earlier in Chukotka.

I sent him an email with a list of questions that would help Luca and me determine the legitimacy of the Florence khatyrkite sample. His initial note was more than I could have ever hoped for:

Уважаемъй профессор Пол Стейнхардт! Благодарю за ваше письмо. Я внимательно слежу за дискуссией вокруг условий образованпя квазикристалов и хатеркита в частности, поэтому хорошо представляю значимость этой находки и с . . .

Which in full text translates to:

Dear Professor Paul Steinhardt!

Thank you for your letter. I am carefully watching the debate about the formation conditions for quasicrystals and khatyrkite, so I can well imagine the significance of this finding and, with pleasure, will help to ensure that you understand the conditions for the formation of khatyrkite. You may use the information I have provided you with no restrictions.

In 1979, I did expeditionary research work in the Russia Academy of Sciences as a researcher for the North-East Scientific Research Institute, Academy of Sciences of the USSR in Magadan. The plan was for a large expedition, but, because of bad weather, the only ones who arrived at the Iomrautvaam River were me and a student from Yakutsk. I conducted the work at the Listvenitovyi Stream, a small stream which is no more than a mile in length and is the right tributary of the Iomrautvaam River. This is a unique stream, despite its small size. For several years it was a place to search for placer gold. The year before my arrival, it was completely depleted. The stream bed was flattened by a bulldozer. As a result, on the left side of the creek was bared a meter-thick layer of blue-green clay, probably a serpentinite by-product, a chemical weathering crust. Washing it was extremely difficult but it could be done with hot water. I washed more than 150 kilograms of material. After washing in a heavy mineral concentrate, this mineral was found that immediately attracted attention. It was a pyramid 4 mm in height and 4 mm at the base. I was attracted by its bright silver color, whiter than native platinum metals, but lighter in weight. Arriving in Magadan, I gave this sample to LV Razin, because he was the team leader for the study of platinum deposits which included me. After a time he told me that this was not a platinum mineral: I had found four new phases of aluminum-based minerals. The next year, Razin left the institute and moved to another city. I never met him again. Years later came the publication of khatyrkite and cupalite as new minerals by Razin.

I think that there has been no geological study of Listvenitovyi Stream except for mine. And, most likely, nobody brought khatyrkite from there. Many questions about the nature of the formation of this clay remain unresolved. It is present in the stream in large quantities. Organization of an expedition to study the creek is possible. From this place to Anadyr is about 200 kilometers. I am very familiar with the area. The stream is not marked on the map you sent me, but I will try to show the place, using images from satellites and send them to you shortly.

Sincerely VV (Valery) Kryachko

Valery's detailed response, along with his answers to my follow-up questions in later emails, proved to me beyond a shadow of a doubt that Valery Kryachko, not Leonid Razin, was the person who had recovered khatyrkite samples from the Listvenitovyi Stream during the summer of 1979.

I finally understood why the Razin paper had been so devoid of details. It wasn't because Razin was trying to fictionalize characters and locations in order to hide his location from competitors. It wasn't because Kryachko was a Chukchi who had disappeared back into the wilderness. And it clearly wasn't because Kryachko was deceased.

I concluded that Razin had written the text based on what he recalled from Valery Kryachko's oral account instead of inviting him to be a coauthor. Perhaps it was because Valery was only a lowly student at the time. For whatever reason, Razin simply did not share the credit with him. It took nearly a quarter of a century after the publication of that paper to discover the truth: Let the record show it was Valery Kryachko who first discovered samples that later proved to have the new crystalline minerals khatyrkite and cupalite and the first natural quasicrystal.

To my surprise, Valery was already familiar with our *Science* paper, published about seven months earlier, announcing the discovery of the first-ever natural quasicrystal. Since it involved a Russian sample, the news had been broadcast on national media.

Until I had contacted him, though, Valery had no idea that he might be personally connected to the natural quasicrystal story. It was my honor to inform him that he was a central figure in the discovery. Valery was elated to hear the news and immediately volunteered to do anything he could to help.

PRINCETON, JANUARY 2010: Later the same month, I met with Lincoln to bring him up to date on the latest series of incredible happenings. I told him the story about how Luca and I had managed to track down Valery Kryachko and that he was indisputably the person who had pulled the St. Petersburg holotype out of the ground, which was very likely the source of the Florence sample as well.

Over the past few months, I learned the hard way that Lincoln never attempts to hide his displeasure. Now, I was about to experience the opposite was true, as well. Once Lincoln heard the good news, it was as if a dark cloud suddenly vanished from the room. I watched a grin come over his face and knew that the legendary Lincoln Hollister was officially on board. It was just the reaction I had been hoping for.

I was still relishing the moment when he made a startling proposition: The next step is an expedition back to Far Eastern Russia to look for more samples. "You simply must go," Lincoln insisted.

Lincoln then wrote to his red team colleague, Glenn MacPherson, at the Smithsonian:

> Paul has come up with stuff that points to the provenance of the sample in the eastern Koryak Mts. He found the guy that collected the original sample, and the connection between the Florence sample and the St. Petersburg sample is persuasive. I think there is enough to support a proposal to NSF to go to the locality and figure out the geological setting of the locality and, with luck, get more sample.

I was not quite sure what to make of it. But I could sense that the search for natural quasicrystals was about to switch gears, and would be moving from the laboratory into a realm I knew nothing about.

FIFTEEN

SOMETHING RARE SURROUNDING SOMETHING IMPOSSIBLE

PASADENA, MARCH 19, 2010: It was a beautiful day in Pasadena and a warm reminder of one of the things I enjoy the most about Southern California—the weather. Patchy snow was still on the ground back home in Princeton. But here, spring was already in full bloom. I soaked up the sun as I headed across the Caltech campus.

I walked along one of the main routes, which took me down the Olive Walk and past a familiar two-story building. Lloyd House, my freshman dorm. I was flooded with memories as I walked by. I had experienced my first-ever earthquake in that dorm, a frightful jolt that knocked me out of bed in the middle of the night. I remembered the awkward moment, also as a freshman, when I screwed up enough courage to offer a timid greeting to my physics hero and eventual mentor, Richard Feynman.

I recalled again his famous admonition: *You are the easiest person to fool.* And all these years later, it was Feynman's advice that brought me back here. I wanted to be absolutely certain that I was not fooling myself about our quasicrystal investigation.

I was heading toward a lunch meeting at the faculty club with Ed Stolper, the highly respected Caltech provost and well-known geologist. Ed had a reputation as a critical thinker who could be unsparingly, perhaps even brutally honest. I was betting on his frank assessment of our investigation. Ed had spent his career studying both natural and man-made materials. Working on JPL's Mars Exploration Rover mission, he had discovered remarkable similarities between a rock on the Martian surface and a rare rock sample found on Earth. He had also made special studies of man-made materials exposed to the elements, weathered pieces of slag that could easily be mistaken for a natural sample. That made his work directly relevant to our investigation. I was armed with a huge binder stuffed with our accumulated research. With his vast expertise, Ed would be able to judge if the sample we were studying was potentially worthwhile, or if there was any possibility it was just an old beat-up piece of scrap metal.

I had reviewed my presentation with Luca. But I had not told him everything. The meeting carried more downside risk than I cared to admit. The truth would have made him sick with worry. Ed was well known and highly respected throughout the scientific community, so any hint of disapproval from him would be disastrous. If he expressed any serious doubts, Lincoln and Glenn would run away from the project. Rumors would spread and eventually undermine my relationships with other respected geologists, whose support was crucial to our research efforts.

Ed and I had never met, but we recognized each other immediately from photographs. He had wavy brown hair, glasses, and a welcoming demeanor that helped put me at ease. We sat down for lunch and got to business fairly quickly.

He listened patiently as I made my lengthy presentation using figures and tables from my thick notebook. By now, our investigation was a complicated story, and I had spent quite a bit of time organizing all of the relevant details.

Ed took note of our theories of how the quasicrystal might have formed. He also considered the fact that many other experts believed the Florence sample must be slag and the evidence that Luca and I had collected suggesting otherwise. He interrupted me occasionally to ask incisive questions but never revealed what he was thinking, not even when I admitted that we had not found any decisive evidence to support our theory that the Florence khatyrkite sample was natural.

Lunch was nearly over by the time I finished laying out our case. I sat back, knowing I had given it my best shot and waited to hear his reaction. *Would this be the end?* I wondered.

Just as I expected, Ed delivered a frank, no-nonsense opinion, beginning with the bottom line. He firmly declared that there was "no chance" that our sample was synthetic. It was definitely not slag, he said, or anthropogenic.

Fantastic! I thought. *The gamble had paid off! I wish Luca were here to share this moment.*

Ed pointed to several of the chemical and geological clues from my presentation to support his conclusion. He also weighed in on the possible theories I had mentioned about the sample's formation: lightning strikes; volcanoes; hydrothermal vents; collisions between tectonic plates; debris from rockets or jets; and, of course, deep Earth processes and meteorites. He thought the meteorite explanation was unlikely and preferred some of the other theories we were still considering.

My pent-up anxiety was slowly melting away as I listened to Ed. It is always hard for me to explain to my university students how difficult it is for a scientist, even an established scientist such as myself, to challenge conventional wisdom. Everything always appears to be so simple to others in retrospect. They lose sight of the fact that making scientific progress is always a struggle that requires a great deal of personal endurance. There is a huge amount of peer pressure to conform. For example, after Luca and I suggested that our sample of metallic aluminum might be of natural origin, which was generally thought to be impossible at the time, we were subjected to more than a year of skepticism and withering criticism from certain experts, including our own colleagues on the red team. It had not been easy. The negative comments were sometimes so harsh that the two of us were left dispirited. But work is a great coping mechanism. We kept plowing ahead, incrementally gathering additional evidence to test our thesis. After fourteen months of hard work, it was greatly satisfying for me to hear Ed validate our efforts.

As the meeting ended and I expressed my gratitude for his time and expertise, Ed left me with one last, and as it would turn out, crucial piece of advice. He suggested that we analyze the abundance of rare oxygen isotopes in our sample. It was a novel line of inquiry that Luca and I had never considered, and something Ed thought would help us reduce the remaining list of possible explanations.

I sat alone at the table for a short while after Ed returned to his office and marveled to myself at what had just happened. I gathered up the pages from my notebook that were strewn about the table and wrote a careful record of the meeting to share with Luca.

Ed had grilled me with penetrating questions. Time after time, I was able to reach into my thick notebook and locate a figure or table that provided a precise, unambiguous answer. Being able to answer all of Ed's questions so completely made me appreciate the wealth of evidence that Luca and I had assembled. Our seemingly scattergun, all-out approach of pursuing every possible lead and conducting every possible test had paid off. Thanks to Ed Stolper, the case for a natural quasicrystal would have to be taken seriously by everyone in the scientific community.

Our investigation is definitely legitimate, I told myself.

I spent the next few hours alone, strolling through campus and the surrounding neighborhood. I reveled in the wondrous arrival of spring and reminisced about Feynman, wondering what he might say.

PRINCETON, LATE MARCH 2010: A few days later, I returned home to winter's cold embrace. Lincoln and Glenn were suitably impressed when I told them about Ed Stolper's analysis and generally positive reaction. It was promising, they conceded. But they continued to grouse that we still did not have decisive evidence that our sample was natural.

FLORENCE, MAY 17, 2010: Six weeks later, the blue team was finally prepared to deliver what the red team was asking for.

Ever since our initial discovery of a natural quasicrystal fifteen months earlier, Luca and I had been grinding away in our separate laboratories, searching for clues among the dwindling number of ever-smaller grains of the Florence sample. A month after my discussion with Ed, Luca began examining a grain that was only 70 nanometers across, about one-hundredth the width of a human hair, when he spotted something truly remarkable. Rather than send me an email explaining what he had found, he made an appointment for us to Skype chat the next day, with the promise of "some news" he wanted to present.

When we connected the next day, Luca typed into the Skype chat box: "Please give me five minutes to prepare the file for you. . . . I have amazing news . . ."

My immediate response was to type: "Ok!!! But I have been on the edge of my seat all night!" Luca was not one to exaggerate test results, so I was anticipating an important development.

"The surprise will be great," Luca typed. "Trust in me."

I sat waiting impatiently for what felt like an eternity. Finally, a large electronic file arrived from Florence. I opened the file, and my eyes went wide as the image appeared on my screen. I caught my breath. I could not believe what I was seeing. Luca had discovered a grain of *stishovite*.

My mind began to reel at the implications. "This is amazing!" I wrote. "How sure is the identification of stishovite?"

"100% plus," he replied.

Stishovite is a famous mineral named after the Russian physicist Sergey Stishov, who first manufactured it in his laboratory in 1961. It can only form at extremely high pressures, about 100,000 times the pressure exerted by the Earth's atmosphere at sea level. A short time after it was discovered in the lab, a natural example of stishovite was discovered at Meteor Crater in Arizona. Upon further study, scientists proved it had formed as the direct result of the hypervelocity impact of a meteor with the Earth.

The discovery of stishovite in the Florence khatyrkite sample supported our opinion that it was, indeed, natural. The pressures needed to create it could never be reached in any industrial process. Stishovite is well known as an indicator of an ultra-high-pressure phenomenon, something far beyond the scope of any of the normal geological processes that occur on the surface of the Earth.

The chemical composition of stishovite is very familiar: SiO_2, silicon dioxide. One part silicon to two parts oxygen, which is the same chemical formula that applies to ordinary sand or window glass. What makes stishovite so distinctive is the way the atoms are arranged. The process is directly analogous to carbon atoms, which form one kind of crystal arrangement on the surface of the Earth, resulting in graphite, and a different arrangement of atoms at high subterranean pressures, resulting in diamonds. Silicon dioxide molecules also make different crystal arrangements depending on whether they are created at ordinary pressures, resulting in sand, or ultra-high pressures, resulting in stishovite.

The difference between stishovite and sand can be unambiguously detected by looking at the spacing and arrangement of sharp Bragg peaks in an electron diffraction pattern. Luca had already performed those tests and sent me a series of diffraction patterns that left no doubt about the sample's identity.

A few days later, Luca emailed me the magnified image of a tiny region of the stishovite grain that provided even more stunning news.

The blurry black-and-white image seen below may appear unimpressive. But from a scientific perspective the photo truly amazes.

The image is a combination of something very rare surrounding something impossible. Stishovite, the silvery material, is the rare substance. It is seen surrounding an icosahedral quasicrystal, the black slug, something once considered to be impossible. Actually, "doubly impossible" would be a more apt description of the quasicrystal. Impossible because of the forbidden five-fold symmetry. Impossible again, because of the chemical composition that included metallic aluminum, which had never been seen to occur naturally.

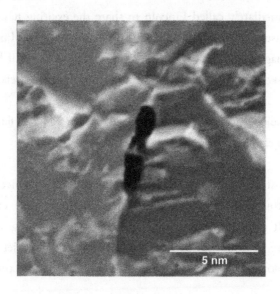

We knew that the stishovite was the product of a high-pressure phenomenon which is only possible under certain circumstances: deep under the surface of the Earth, during a collision in outer space, or as a result of the impact of a very large meteorite with the Earth's surface. The pressures involved were far above what could be reached by any normal anthropogenic activity.

In this particular sample's case, we could immediately rule out the possibility that the quasicrystal formed when a large meteorite struck the Earth's surface. That would have melted the aluminum-rich metal seen throughout the Florence sample and caused it to chemically react with the oxygen in the Earth's atmosphere.

The fact that the quasicrystal survived the high pressures required to make stishovite taught us another lesson. It told us that the quasi-crystal already existed when the stishovite was formed, and that to-gether they had managed to endure the kind of ultra-high pressures that can only be found far outside the influence of human activity.

Here was the direct proof of a natural origin that Lincoln and Glenn had been seeking.

I immediately called Lincoln to share the exciting news. I took great pleasure composing the email to Glenn, attaching Luca's most recent image along with our analysis. I was eager to see if he would respond skeptically, as usual. He wound up hedging his bets:

> If it really is stishovite, and there really is QC inside of it, this is a *game changer*.

Almost all our previous theories of how the Florence samples formed could now be discarded. The Florence sample could not be slag. It could not have been made by miners fooling around at a campfire. It could not be the result of exhaust from a jet airplane. It could not have been made by an explosion or manufactured in an ordinary laboratory. It could not be caused by lightning, or by hydrothermal vents or volcanoes. None of the theories others had postulated could produce the extraordinary pressures required to form stishovite.

Equally revealing was the way the stishovite and the quasicrystal were fused together. This proved that quasicrystals are not as fragile as previously supposed. Since the quasicrystal was totally enclosed in the stishovite grain, it meant the quasicrystal could survive ultra-high pressures.

Glenn conceded all these points but wanted to exhaust every conceivable alternate explanation, no matter how remote. As a last gasp, he asked us to consider whether the sample might be the result of an atomic bomb test. Luca and I were easily able to eliminate that idea because measurements showed the sample did not have any of the heavy elements that would be the by-products of a nuclear explosion.

There were only two plausible theories left that could explain the presence of stishovite. The sample could have been created in inner space, formed thousands of miles below the Earth's surface and conveyed to the outer crust in a superplume. Or it might be a visitor from outer space, a fragment created by the violent collision of two asteroids.

Which possibility was right? And how could we prove it?

SIXTEEN

ICOSAHEDRITE

PASADENA, MAY 2010: *Inner space or outer space?* That was the question.

In my mind, the meteorite theory, outer space, was always the leading explanation for the origin of our natural quasicrystals. Meteorites contain a much wider variety of metals and metal alloys than terrestrial minerals. But we needed something more than a rational argument. We needed incontrovertible evidence.

Two months earlier, Ed Stolper had advised me that we could decide the issue by analyzing the abundance of rare oxygen isotopes in our sample. He referred me to geochemist John Eiler, who studied the origin and evolution of meteorites, among other things. But until we discovered the stishovite sample, I did not feel confident enough to impose on John and ask him to use his expensive equipment to test our sample.

John worked closely with Yunbin Guan, the director of Caltech's Microanalysis Center. The Center had a valuable piece of equipment called a NanoSIMS, a nanometer-scale secondary ion mass spectrometer, which could perform the oxygen isotope test that Ed Stolper had recommended. There are only a handful of NanoSIMS in the world, and only a few institutions, including Caltech, allow outsiders to collaborate on projects.

John had agreed to use the NanoSIMS to perform the oxygen-isotope test on one of our tiny grains. I flew to Pasadena to meet with him, carrying our tiny grains in a small sealed box carefully tucked in my book bag. There was no way that precious box was going to be put in checked luggage.

Once I arrived at Caltech, John and I spent several hours reviewing all the measurements made by our team in our three labs: Princeton University, the University of Florence, and the Smithsonian Museum of Natural History. It was the same set of data I had shown Ed two months earlier, coupled with our recent discovery of stishovite.

Like Ed, John concluded that the likely origin of our sample was terrestrial. "Aspects of the olivine mineral grains remind me of terrestrial samples I have studied," he said. Although my inclination was that our sample came from outer space, I greatly respected Ed and John's opinions and kept an open mind.

I was truly delighted that John was eager to help. He was smart, energetic, and precise. But I was deeply disappointed by what he told me next. The NanoSIMS was a mercurial piece of equipment and was currently under repair. It would probably be a few months before it would be back in working order.

Luca and I had no choice but to wait. The next few months proved to be agonizing. Every day Luca and I had a Skype exchange during which we debated the possible outcome. Inner space or outer space? Outer space or inner space? But what if the test was inconsistent with either a terrestrial or an extraterrestrial origin? Then, despite all our precautions, we would have to consider the dreadful third possibility. Malicious mischief.

The NanoSIMS is an unforgiving test that can unmask a fake material in a heartbeat by measuring the distribution of nuclear isotopes in the sample. This was a feature that a hoaxer would not have anticipated and could not have faked. The longer we had to wait for the test, the harder it was to get this depressing possibility out of our mind.

PASADENA, JULY 2010: Two months later, we were told that the NanoSIMS was finally back in working order and our sample would be tested sometime during the last ten days of the month. Every few days, I would check to see if the test had been performed. "Not yet" was always the reply. The excruciating wait continued.

The NanoSIMS was an ideal instrument to study the isotopes of oxygen atoms in our sample. Atoms are distinguished from one another by the number of protons. For example, all oxygen atoms have 8 protons, which is why it is listed as the eighth element in the Periodic Table. The term "isotope" refers to atoms with the same number of protons, but with different numbers of neutrons.

There are three stable types of oxygen isotopes, each with 8 protons but different numbers of neutrons. The most common type of oxygen atom has a matching number of protons and neutrons: 8 protons and 8 neutrons. Since $8 + 8 = 16$, this isotope is labeled ^{16}O. But there are two less common oxygen isotopes. ^{17}O has 9 neutrons. ^{18}O has 10 neutrons.

If you were to study all of the oxygen atoms in the air you breathe, you would find that 99.76 percent of them are ^{16}O, 0.04 percent of them are ^{17}O, and 0.2 percent of them are ^{18}O.

The relative percentages on Earth are determined by the history of the planet and the exposure of its minerals to cosmic rays and radioactivity. Other planets, like Mars, have had different evolutionary histories and their minerals have been exposed to different levels of cosmic rays and radioactivity. So minerals from Mars contain percentages of the three oxygen isotopes that are different from those found on Earth. The same applies for minerals formed on other planets and on different types of asteroids.

By using a NanoSIMS to measure the amounts of the three oxygen isotopes in different minerals in a sample, a geochemist can determine if the sample is natural and, if so, where it came from.

PASADENA AND PRINCETON, JULY 26, 2010: At last, the long-awaited email arrived from John Eiler announcing that the NanoSIMS measurement had been performed and the results analyzed:

> The two important findings are: the ^{17}O anomaly is significantly less than 0; and the ^{18}O is very low.

I felt like screaming out of sheer frustration because, after all these months of waiting, I had no idea what that sentence signified. It was geochemist-speak, and I am not a geochemist. As I continued reading, I was happy to discover that John had translated the findings into a language that I could understand, using a plot similar to the one on the next page to illustrate his point.

The horizontal axis of the graph represents the amount of the rare isotope ^{18}O found in the sample compared to the most common isotope ^{16}O, which is known as the ^{18}O anomaly. The vertical axis represents the amount of the rare isotope ^{17}O compared to the most common isotope ^{16}O, which is known as the ^{17}O anomaly.

Also shown are two grayish lines that meet toward the upper right. The two lines cross at a point that roughly corresponds to levels one would measure in ocean water on the Earth. The upper line labeled TF refers to Terrestrial Fractionation, and indicates what fractions of ^{17}O and ^{18}O are found in various kinds of minerals that formed on the Earth. Because rocks on Earth form in different ways, they do not have exactly the same isotope distribution as ocean water, but rather, can take any of the values along the TF line.

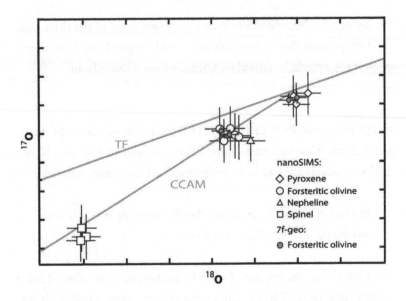

The circles, diamonds, squares, and triangles with lines sticking out represent the values measured for different types of minerals found in the Florence khatyrkite sample: pyroxene, forsteritic olivine, nepheline, and spinel. They do not lie along the line labeled TF, terrestrial fractionation, which indicated that the Florence sample did not originate anywhere on our planet.

Equally important, the results were not randomly distributed, as they might have been if the material was an intentional hoax or had been accidentally synthesized in a laboratory or aluminum foundry. Instead, all the data aligned along a different line labeled CCAM.

CCAM is the acronym for Carbonaceous Chondrite Anhydrous Mineral, a technical term for a stunning conclusion. It was the geochemist's way of saying that the Florence sample, including our quasicrystal, was definitely an extraterrestrial. A visitor from outer space. A meteorite.

More specifically, a CCAM is a rare type of meteorite called a "CV3 carbonaceous chondrite."

Luca and I were all too familiar with CV3 carbonaceous chondrites, especially the most famous one called Allende. The Allende meteorite had nearly derailed our project.

One year earlier, Glenn had concluded that the powdery material in the vial marked 4061-Khatyrkite from Curzio Cipriani's home laboratory was actually from the Allende meteorite. He had memorably ascribed the mix-up to a "capricious, if not overtly malicious, God." From that, he had gone on to conclude that Cipriani was careless and the mineral collection from Florence untrustworthy. The fallout nearly prevented the initial public announcement of our discovery of a natural quasicrystal.

As a result of the NanoSIMS test, Luca and I now knew that Glenn was wrong to identify the material as Allende. But there was a perfectly good reason for his error. The Florence khatyrkite sample was exactly the same type of rare meteorite. Both were created 4.5 billion years ago at the beginning of the solar system under similar conditions and contained many of the same minerals. No wonder Glenn had mistaken one for the other.

But they were not an exact match. The Florence sample was even more intriguing than Allende because it contained aluminum-copper metal alloys that had never been seen before in any other known rock or mineral. Therefore, it was arguably more important than Allende, because it contained evidence of physical processes in outer space that had been previously unknown. Those processes might have affected the evolution of planets and the earliest stages of our solar system. *But how so?*

Luca and I knew where to go for the answer. Right back to one of the world's foremost experts on CV3 carbonaceous chondrites: Glenn MacPherson. Glenn had been criticizing our work for the last year and a half. He was the strongest advocate for the theory that the Florence sample was a worthless piece of slag. From the very moment he and I had first laid eyes on each other on the steps in front of the Smithsonian Natural History Museum, Glenn had been explaining to me why, in his expert opinion, the sample could not possibly be a meteorite.

The original images recovered from Luca's damaged hard drive had only indicated a jumbled mess of a dog's breakfast, as far as Glenn was concerned. Neither the Amsterdam detective story that led to a connection with the St. Petersburg holotype nor our eventual contact with Valery Kryachko, who had originally recovered the material from a Chukotka dig site, had seriously altered his view. Glenn even managed to retain some degree of skepticism after Luca discovered stishovite enclosing a slug of quasicrystal in the sample.

The NanoSIMS test could have invalidated all of our hard work and proven Glenn's skepticism to be well-founded. Instead, it produced the opposite result. It instantly confirmed our original hypothesis that the Florence sample was natural.

To Glenn's credit, it also eliminated every last fiber of resistance. Always known for a clever turn of phrase, Glenn's email response after receiving my report of the NanoSIMS results began with a simple subject line: "Welcome to my world."

First, congratulations—you have an extraterrestrial sample. I work with oxygen isotopes all the time, so I understand the diagram extremely well and what it implies. . . . This new data changes everything. You can call off your expedition to Siberia for one thing, and you can stop worrying/thinking about things like ultrahigh pressures and lower mantle and serpentinization and all that other stuff. . . .

But now we have several new mysteries. If this thing really did come out of a sediment deposit and it really is a meteorite, I don't know how they found it and I don't know how it survived. . . .

This whole project is now suddenly and squarely in my realm, which means that I will have to take a more central role in guiding it. *Welcome to my world.*

Convincing Glenn was a major milestone. Luca and I had enormous respect for his expertise and his intellectual honesty. Lincoln Hollister was also delighted to learn of Glenn's reaction. Both members of the red team happily conceded to the blue. We were now unanimous.

Luca and I could not help but be amused, though, with Glenn's sudden claim of ownership. Of course, we had absolutely no intention of giving up our leadership roles.

PRINCETON AND FLORENCE, OCTOBER 1, 2010: Two months after the confirmation from Caltech's NanoSIMS test, Luca sent me even more good news.

The International Mineralogical Association Commission on New Minerals, Nomenclature and Classification had just voted to accept our quasicrystal as a natural mineral. They also accepted our proposed

---

**INTERNATIONAL MINERALOGICAL ASSOCIATION
COMMISSION ON NEW MINERALS, NOMENCLATURE
AND CLASSIFICATION**

Chairman: Professor Peter A. Williams
School of Natural Sciences
University of Western Sydney

Phone: +61 2 9685 9977
Fax: +61 2 9685 9915
E-mail: p.williams@uws.edu.au

Postal address: School of Natural Sciences, University of Western Sydney, Locked Bag 1797, Penrith South DC NSW 1797, Australia

1 October, 2010

Dear Luca,

Congratulations on your new mineral, icosahedrite (2010-042)!

name: "icosahedrite," a fitting name for the first known mineral with icosahedral symmetry to be entered into the official catalog.

I savored the historic moment. This was one of the milestones I had been seeking since first imagining the possibility of natural quasi-crystals. But I knew the story was not yet over. I went back to reread Glenn's message in which he wrote,

You can call off your expedition to Siberia for one thing . . .

I stared at the note and shook my head. *He's definitely wrong*, I thought.

There was now convincing evidence that our sample was a visitor from outer space and most likely a creation dating back to the birth of the solar system. But many mysteries remained. How did it first form? Why did it contain quasicrystals? What path did it take through space before entering the Earth's atmosphere? How did pieces of it become lodged in the blue-green clay of the Listvenitovyi? And why had it not corroded since its arrival on Earth?

The few remaining specks of the Florence sample were not enough to answer any of these questions. We needed to recover more material from the same source. The only way to resolve the remaining mysteries was to charge ahead with an expedition to Far Eastern Russia in search of more specimens. That was absolutely clear in my mind.

What I never envisioned, though, was that I might be forced to take part in such an adventure.

PART III

KAMCHATKA OR BUST

SEVENTEEN

LOST

MIDDLE OF NOWHERE, NORTH OF KAMCHATKA, JULY 22, 2011:
Somehow the unimaginable had happened. No one was less suited to
take part in, much less lead, an expedition to the remote regions of Far
Eastern Russia. Yet here I was.

For the last sixteen hours, I
had been riding aboard a giant
tracked vehicle that was carrying me, along with half of my
team, across the desolate tundra
of Far Eastern Russia. At last,
close to midnight, we rumbled
down a steep hill and stopped
for the night by a riverbed. I
felt calm and settled despite
the strange and unfamiliar surroundings.

That equilibrium was shattered the moment I climbed
out of the cab and leapt off the
huge tractor treads to the ground

below. Suddenly I was suffocating. Sensing my breath, hordes of mosquitoes had sprung out of the muck and formed a thick cloud around my head, effectively cutting off my air supply.

I was desperate and embarrassed at the same time. Desperate to breathe, and embarrassed to display any vulnerability because I was the supposed leader of this expedition. I tried to hide my distress from the rest of the team by slowly edging away toward the hill behind us, all the while futilely attempting to bat away my attackers.

Desperate to regain my composure, I forced myself to focus on the series of events that had somehow convinced me to travel thousands of miles to this hellish, forsaken, mosquito-infested place.

I was originally suckered into the trip by the discovery of a natural quasicrystal in an unremarkable mineral sample that had been hidden for years in the storage room of an Italian museum. Through extraordinary detective work, we had proven that the sample was part of an ancient meteor dating back 4.5 billion years to a time before the birth of the solar system. Our extraterrestrial had survived numerous interstellar encounters and collisions while hurtling through space at nearly 100,000 miles per hour. It finally entered the Earth's atmosphere about seven thousand years ago, around the same time human beings were inventing the wheel. The meteor announced its arrival by heating to a spectacular incandescence, streaking across the sky just south of the Arctic Circle before landing in the Koryak mountain range just north of Kamchatka, where it remained undisturbed for thousands of years.

In 1979, a young Russian student named Valery Kryachko was hired to search for platinum along a stream bed in the middle of the Koryaks and accidentally found a piece of our extraterrestrial buried in layers of a mysterious blue-green clay. From there, the unrecognized sample began a thirty-year journey which eventually brought it to our laboratories, where we discovered natural quasicrystals embedded inside.

Now, several years after that discovery, we were returning to the Koryak Mountains with that same Russian student, now sixty-two years old, in search of more pieces of the meteorite.

We knew the Florence museum sample was controversial. The weird chemical compositions of the quasicrystals and some of the metallic crystal alloys we identified contradicted scientific principles about what forms of matter could exist in nature. So in spite of all the evidence we had gathered, a few scientists continued to question whether the sample was natural at all.

We had been chipping away at the Florence sample for the past two years on two continents, performing every test imaginable to learn the secret behind the never-before-seen minerals. We had conducted so many tests, in fact, that we had managed to destroy all of the sample and had nothing left to study. The only way to advance the science was to travel to the Koryaks and find more material.

Having come this close to proving natural quasicrystals exist, I had no choice. I was absolutely compelled to go! Wasn't I?

Although I was still suffocating, barely squeezing in quick breaths of air while swatting away at the insect swarm, these inner thoughts began to calm me down.

And then, a tsunami of panic swept over me as a truly terrifying thought came to mind. *The second vehicle carrying the other half our team—including my youngest son Will—was missing.*

The two vehicles had been traveling together most of the day, but Will's truck had been limping along because of mechanical problems and had slipped out of sight a few hours earlier. Our driver had assumed it was trailing just beyond our view. But if that were the case, it should have caught up with us by now.

Standing alone at the top of the hill with my stomach in knots and my mind reeling, desperate for any sign of Will and the rest of my team, I felt an overwhelming sense of guilt. I had challenged the impossible once too often with the horrifying result of having put my own son's life in danger and perhaps losing half of my team.

One by one, I had blown past explicit warnings from all kinds of experts: *Searching for new samples is hopeless. A trip to such a remote region is not worth the risk. The expedition will never be fundable. Pulling together a team in just a few months is unachievable. You took a wrong turn in this investigation. It's foolhardy for someone like you to make this trip. That expedition site is inaccessible.*

Hopeless. Several geologists told me the chance of finding another piece of our meteorite in the vast emptiness of Kamchatka was less likely than pinpointing a needle in the proverbial haystack.

Have you looked at a map, Paul? Talk about impractical. You don't even have the GPS coordinates of the original dig site, do you?

Colleagues warned me that I was putting too much faith in a sixty-two-year-old Russian scientist and his ability to remember crucial details of a long-ago incident that had no special significance for him at the time. Even if Valery Kryachko was capable of leading us to the exact spot where he had discovered grains from the meteorite more than thirty years ago, which everyone considered unlikely, the chance of finding additional material from the same meteorite was infinitesimal because the grains would be tiny, thinly dispersed, and hard to discern from the millions of other grains in the area. No geologist in their right mind would commit time and resources to such a crazy endeavor.

I had to concede that our chances of success were close to nil. But at the same time, they were not zero. And as long as there was a nonzero chance of finding more samples and solving the mystery of their origin, there was no choice but to pursue and persist. My timing was dictated by the fact that Valery, the only person in the world who could point us to the source, was willing and able to go. It was now or never. Or so I convinced myself.

Not worth the risk; quit while you're ahead was the message from several influential physicists.

Luca Bindi and I should rest on our laurels, they argued. Our *Science* article had already convinced nearly all of the scientific community that natural quasicrystals exist. Why take the chance of going to Russia and coming up empty or, worse, finding confounding evidence? Either result might further encourage the few skeptics who still question your conclusions and could raise doubts about your entire investigation.

I understood that a failed expedition could put our credibility at risk. But I had been searching for supposedly impossible quasicrystals for nearly three decades, and as a result had become somewhat immune to the skepticism of other scientists. My Italian colleague, Luca Bindi, felt the same and our shared obstinacy had taken us down a number of blind alleys. But it had also enabled us to make some amazing discoveries and I was not prepared to put an end to that process. We could not let the fear of failure prevent us from doing our utmost to solve the scientific mysteries that remained.

Not fundable. Where would I find enough money for such a preposterous expedition? was a fairly universal reaction.

Funding agencies would never provide financial support based on a convoluted detective story, the thirty-year-old memory of some unknown Russian mineralogist, and a few microscopic grains of material. The risk of failure was much too high.

Just as everyone predicted, I received strongly negative responses from the National Science Foundation, the Department of Energy, the American Museum of Natural History, the Smithsonian, the National Geographical Society, and other notable funding agencies. Don't even bother making a formal proposal, they cautioned me.

I had not expected a traditional funding agency to consider my request, so I was prepared for all of the unfavorable reactions. I was also prepared when my own institution, Princeton University, declined to help. I knew the university had many other priorities and was generally more focused on less risky projects that directly benefited students on campus.

My only hope was to find a wealthy and generous benefactor. But before I could search for such a person, I had to get consent to do so. Princeton, like most American universities, forbids faculty members from soliciting private monies because it might interfere with the university's fund-raising efforts.

Knowing this, I asked if I could proceed under the strict proviso that I prove my benefactor would never consider donating to Princeton University for any purpose other than supporting my expedition. Finding someone willing to contribute to a Princeton-led project who otherwise had no interest in Princeton was impossible, by their reckoning. Perhaps that was why they allowed me to give it a try.

It must have been a surprise when I called the Development Office three days later to report that I had identified a candidate. I had found a wonderful donor with no connection to Princeton who was willing to provide me with the $50,000 that I needed for the expedition. After several weeks of investigation, administrators agreed that I had, to their admitted astonishment, complied with the terms of the deal. The donation was made to the university, earmarked for my expedition.

Months later, when our projected costs soared because of a significant change in the transportation plan, I was forced to return to my donor to ask for more support. I was overwhelmed by his reaction. Without a moment's hesitation, he graciously agreed to cover the overage, which was more than double my initial estimate.

I remain impressed by my benefactor's remarkable modesty and his continued insistence on remaining anonymous. I gratefully refer to him as "Dave," a true friend of science.

Unachievable. Why are you rushing into this? was the reaction from every geologist I spoke to with experience working in Russia.

This whole thing is much more complicated than you think, Paul. Your plan is unrealistic. You will never be able to get a team of qualified people together in such a short amount of time. Also, Kamchatka is a restricted area—don't you know you need a series of high-level approvals from the Russian government to get near that place? It's impossible to speed up that process. *What a hare-brained scheme!*

These were fair criticisms. I needed to recruit a highly skilled team that was ready to drop whatever plans they might have to join the expedition. Ten months did not sound like enough time to negotiate the famously byzantine Russian bureaucracy. We would have to obtain permission from the Russian government in Moscow, the regional government in Chukotka, the Russian military, as well as the FSB, the Russian security agency. Visitors to the region are closely scrutinized because of Russia's historic emphasis on its strategic importance.

On top of all that, we would have to organize everyone's travel to Anadyr, the Russian town nearest the dig site with a substantial airport. We would then have to line up weeks of food supplies, gather equipment, and figure out how to transport our team from Anadyr to our remote destination at the Listvenitovyi Stream.

One step at a time, I told myself. My first challenge was to find experts willing to sign up for an expedition that many of their peers thought was ill-advised.

I immediately emailed the Russian geologist Valery Kryachko. He was the most important member of the team because he was the only one who knew how to find ground zero, the original source of the Florence sample. A sturdy man with a full white beard, Valery had years of field experience and was extremely capable, resourceful, and self-reliant. Fortunately, he not only agreed to go but was willing to do anything in his power to make the expedition a success.

I also reached out to Vadim Distler, who had been Valery's PhD advisor decades earlier. I knew the two of them had teamed up over the years for a number of expeditions to Far Eastern Russia in pursuit of valuable ores and minerals. Warm and friendly, Vadim was in his eighties and an unrepentant chain-smoker.

Vadim recruited Marina Yudovskaya, an experienced field geologist with a razor-sharp mind who had recently succeeded him as head of his division at the Moscow mining institute. She was from Kazakhstan and the daughter of two geologists. Tall, blond, and lanky with a relaxed, easy smile, she was the only member of the Russian contingent who spoke fluent English.

The three Russians knew each other and had worked together for many years. On the other hand, I would have to build the rest of the team from scratch.

Luca Bindi was a must. It had been more than two years since he had discovered the original sample stashed away in a back room of his museum. Luca was the only person who had been able to examine the material in its original state, before it was sliced for testing and pulverized beyond all recognition. His experience with the initial sample, plus his keen eyes and manual dexterity, made Luca our best hope for identifying additional meteorite grains at the Listvenitovyi.

To say that Luca was hesitant to join the expedition would be putting it mildly. He was definitely not a field geologist, Luca reminded me. So I had to do more than a little arm-twisting to convince him to go. I was glad when he ultimately agreed because I trusted his judgment and by now we had become great friends. The trip would give us another chance to work closely together on what had become a wild investigation.

With a lifetime of geological expeditions under his belt, my Princeton colleague and former red team member Lincoln Hollister was also on my wish list. Lincoln very much wanted to go and would have been an extremely valuable member of the team. He had years of experience dealing with difficult field conditions. At the time, however, Lincoln also had an acute, but temporary, medical issue and decided there was too much of a risk that he might experience a medical emergency while stranded in a remote location where medical help was not readily available. He was the first to point out that an incident like that could endanger the entire expedition. Disappointed at having to withdraw, Lincoln agreed to advise and help plan the expedition. He recruited three candidates that he could personally vouch for: Glenn MacPherson, Chris Andronicos, and Mike Eddy.

Glenn was an obvious choice. One of our former red team members, he was intimately familiar with the investigation. It seemed fitting to recruit our former red team "adversary" as the field team's meteorite expert. No one was more qualified to scrutinize and certify our work than Glenn.

Chris Andronicos had obtained his graduate degree at Princeton with Lincoln as his PhD advisor. Chris was an expert on the powerful geological forces that produce mountains, create earthquakes, establish fault lines, and bend and break rocks. He also had extensive experience in different parts of the world, including the Coast Mountains of British Columbia, the Al Hajar Mountains of Oman, and other areas with rocks and formations similar to what we expected to find on our expedition. I found Chris to be a thoughtful, articulate scientist with wide-ranging expertise and a highly creative mind. With a broad, powerful stature and a lifetime of outdoor experience, he was totally at home in the wilderness.

Lincoln's third nominee was Mike Eddy, an outstanding student set to graduate that year from Princeton in geoscience and begin his doctoral program at MIT in the fall. Mike was thin and muscular, a natural athlete who had set several track-and-field records as an undergraduate. He had spent the previous year doing geological field work in the Aleutian Islands off the Alaskan Peninsula, not far from where we were headed. In addition to his brain power, we would be counting on him to provide some of the muscle we were going to need at the dig sites.

Glenn MacPherson and Mike Eddy were both quick to agree to my invitation. But Chris Andronicos was hesitant. He was pursuing extensive field work that summer in the Sultanate of Oman and an unexpected detour to Far Eastern Russia would interrupt his research. But that was not his only concern about joining the expedition.

"I have to be honest," Chris told me. "I am skeptical about the meteorite theory. If I were to join, it would be to map the local geology in order to look for evidence of a terrestrial explanation, such as superplumes or other unusual geological conditions that might better explain how the Florence sample formed."

Chris knew he was challenging my view that the sample was a meteorite. But having spoken to Lincoln, he also knew that the meteorite theory was incomplete. The presence of copper-aluminum alloys had yet to be explained. He felt alternative ideas, like the superplume theory, deserved further consideration. After all, no one knew what an object brought up by a superplume might consist of, since no one had ever seen material that had originated from near the Earth's molten core. And perhaps we would discover other potential terrestrial sources aside from superplumes to account for the Florence sample if we went to Far Eastern Russia.

"However, I seriously doubt," Chris said, "that you would be interested in bringing someone along on your expedition who is skeptical about your meteorite theory."

I had to laugh because I felt exactly the opposite. He was just the kind of person I wanted on the team. I explained to Chris about how I had worked closely with Lincoln and Glenn for two full years, even while they both vehemently opposed our theory that the quasicrystal discovered in the Florence sample was natural. I deeply value the friendly tension that exists between evenly matched red and blue teams because I have always found it to be the best means of arriving at the scientific truth.

"Opposing points of view are always welcome and strongly encouraged," I assured him. And with that understanding, Chris gladly accepted my invitation.

With an international team of scientists now assembled, I knew we would be needing a Russian translator. As director of Princeton's Center for Theoretical Science, I was acquainted with several postdoctoral fellows from Russia, one of whom nominated a former classmate named Alexander Kostin. The two of them had studied physics together as undergraduates at the prestigious Moscow Institute of Physics and Technology. Alexander, who goes by his Russian nickname, Sasha, was working in Texas as a petrophysicist for an oil company. As it turned out, he had long dreamed of going to the Russian Far East and was eager to join the expedition, even though it meant sacrificing time with his family during their summer vacation in Moscow.

Finally, I decided to add more muscle to the team by recruiting someone I had known since he was a baby. My son Will. His father had never introduced him to the great outdoors, but as a geophysics major at Caltech, Will had gained considerable field experience working under difficult conditions in the Mojave Desert and the White Mountains of California. He had grown considerably taller than his father, and was certainly more athletic. Following the trip, Will would begin his doctoral program at Harvard.

I was thankful that Will knew what to expect in the field and could help me prepare both mentally and physically for the trip. I was less sure about the price I had to pay for his advice. Suddenly, Will had the freedom to set the rules and tell his inexperienced father exactly what to do and when to do it. I knew he was going to get endless amusement from the role reversal.

With the international team in place, our Russian colleagues took over the planning. Marina and Vadim knew how to function within the Russian bureaucracy. They spent months working tirelessly on the required paperwork, ultimately generating a foot-high stack of documents. They prepared, submitted, and pushed through a mind-numbing number of forms and letters. The fact that they managed to accomplish all of it in such a short amount of time was a monument to their experience, professionalism, and commitment to our project.

In the meantime, Valery presented me with a meticulously detailed list of supplies that would be needed for the expedition. Someone would have to go to Chukotka in advance, he said, in order to reserve all of the supplies and book the local transportation. Experienced with the area, he volunteered to make the trip.

In the end, contrary to what all the naysayers predicted, we managed to recruit an expert team and complete all of our planning and preparation in *less* than ten months. The early success may have lulled me into a false sense of confidence, leading me to believe that the rest of our so-called problems would be just as easy to solve. What a foolish assumption that had been.

———————

You took a wrong turn in this investigation. Glenn MacPherson shocked everyone by uttering words to that effect at a critical planning meeting. He almost killed the entire project.

A few months before our journey was set to begin, I invited our Russian colleagues to Princeton to finish some of the organizational planning and to share scientific information with the rest of the team.

Valery had just finished telling the assembled group about how he had dug a sample out of Chukotka's distinctive blue-green clay in 1979. It was shiny and almost entirely metallic, he said, just like the one in the St. Petersburg Mining Museum. Shortly after he returned home from the expedition, he handed the sample to Leonid Razin, the Russian scientist who had sent him to search for platinum. That was the last he ever saw or heard of the sample, Valery said, until decades later, when he received an email from me explaining that it had been the source of the first-known natural quasicrystal. He was understandably ecstatic to learn that he was personally connected to the story.

Razin had apparently taken the sample back to St. Petersburg for testing and eventually discovered that it included two new types of crystal minerals, khatyrkite and cupalite. He had published those findings without ever notifying Valery. As part of the official process of laying claim to discovering a new mineral, Razin submitted a piece to be held in perpetuity at the St. Petersburg Mining Museum.

At that point in the story Glenn, who had been listening quietly, suddenly spoke up. Loudly.

"The St. Petersburg holotype that Razin submitted may very well match Valery's recollection of a 'shiny and metallic' material," he roared, "but the Florence sample, the source of the natural quasicrystal and the motivation for this entire expedition, most definitely does not! In fact, it is the complete opposite! The Florence sample appears to be dull and non-metallic."

Glenn had assumed all along that the two samples had been

physically connected and that Razin had broken them apart so he could put one piece in the museum and keep the other piece for himself. But according to Valery's description of events, that was a false assumption.

"Valery's story means the two samples were never attached," Glenn said. "If they were never attached, what evidence do we have that they came from the same place? . . . and if there is no such evidence, is there any justification for going on the expedition?"

Everyone at the meeting held their breath. This was supposed to be a planning meeting. But without warning, the scientific basis for the entire trip was being called into question. Sensing the tension in the room, Valery listened carefully as Marina translated Glenn's provocative remarks and stood to address the group.

"Let me be firm about this," he said, with Marina translating. "It is true that the sample described in Razin's paper and displayed in photographs from the St. Petersburg Mining Museum matches my memory perfectly. It is just like the shiny grain I found in the blue-green clay of the Listvenitovyi. Before I came here, I, too, was puzzled by the difference between the St. Petersburg and Florence samples."

Valery paused as he turned to look at Glenn. "But I disagree with your conclusion. Earlier today, I learned about the history of the sample. Paul and Luca traced it back to the mineral collector in Amsterdam, then to the smuggler in Romania who obtained it from the lab that was working with my former boss, Leonid Razin. Suddenly, I realized the solution to the puzzle.

"The fact is that I did not find just one sample," Valery explained. "I found several different samples at the Listvenitovyi and gave them *all* to Razin. I never mentioned the others to anyone before now because I never paid them much attention. They were not as shiny because the metal was covered in part by other minerals."

What? More than one sample? My mind raced ahead, knowing where Valery's argument would lead.

"Now think about it," Valery continued. "It cannot be a coincidence that the Florence sample has been traced to the same lab, and the same person in that lab, who was also working with the St. Petersburg holotype. The two samples must be related, which can only mean that the Florence sample, just like the St. Petersburg holotype, came from the Listvenitovyi Stream.

"If that is the case," Valery continued, "we now know there were at least *two* different rocks collected from the same clay that *both* have khatyrkite and cupalite. And if there are two, there are probably more to be found."

Valery smiled. "So, the answer to Glenn's question—'Is there any justification for going?'—is definitely Yes! Absolutely! There is now more reason than ever to go."

Even our resident skeptic, Glenn MacPherson, had to agree, and the expedition was back on course. At the time, it seemed like a moment of triumph.

Months later, standing alone on a hill, facing a howling wind and fearing for the safety of my son and the rest of the team, it occurred to me that I might not be in my present predicament if Valery had not so eloquently won that debate.

Foolhardy. Are you crazy? Whom are you trying to kid? was the general reaction from well-meaning family and friends.

They knew I had never laced up a pair of hiking boots, never built a campfire, and never had any reason whatsoever to crawl into a sleeping bag. In other words, they all knew that I had never been camping before in my life and were flabbergasted that I was about to do so in one of the remotest and least explored places in the world.

At first, I was able to laugh off their concerns because I had no intention of going into the field or working at the dig site. I thought

I could stay in Princeton and monitor the trip over the Internet. My plan was to persuade expert geologists to travel in my place.

Lincoln Hollister, who was helping me organize the trip, firmly nixed that idea. I was the team leader, he reminded me. I had to travel with the rest of the team.

Trying to appease Lincoln, I came up with an alternate plan that was nearly as good as staying home and that would still keep me out of the field. We could set up a communications station in Anadyr, the capital of the Chukotka Okrug and the only town in the region near an airport. Luca and I, along with our translator Sasha, would stay in town while the rest of the team was ferried out to the Listvenitovyi Stream by helicopter. Because of the limited space, seats would be reserved for team members with essential technical skills. There would not be any room on the helicopter for a theoretical physicist like me even if I wanted to go, which I most fervently did not.

It was Valery who would ultimately spoil that plan.

Inaccessible. Soon after he arrived in Anadyr on a scouting mission to reserve food and equipment, Valery sent a message to me through Marina: "Helicopters are no good." We would have to take a different approach.

Helicopters would only be allowed to fly if the weather was ideal, which was impossible to predict in that region. The weather was so changeable that no schedule could ever be relied upon. In fact, helicopter use in that part of the world was deemed so unpredictable that no insurance policy would cover the cost of cancellation or injury.

Furthermore, Valery said, there were very few of them available for rent. Russian oil and mining companies drilling were paying exorbitant fees to keep all the helicopters in the area on standby so they were always available for last-minute needs.

But all was not lost, Valery assured me. He had reserved two large "trucks." And the good news, he announced, was that the trucks would be large enough to carry everyone, including Luca, Sasha, and me. No one would have to stay behind in a nearby town, after all. The new plan appealed to Valery, who had been trying to encourage me to join the rest of the team in the field.

Wait a minute, wait a minute, wait . . . a . . . minute! I thought, as I read the email. *Hadn't Valery gotten the hint that I have been trying to find a way to avoid taking part in this expedition? Doesn't he realize that I have never spent a single night in the wilderness?*

I hurriedly examined all my maps and satellite photos of the area to try to figure out which route Valery's trucks would be taking. Lincoln and I had never found any roads on the maps before, and they did not magically appear now. When I asked Valery about the lack of roads, though, he told me not to worry. There are roads, he assured me, or at least that was the way his cryptic message was translated.

So to the dismay of family and friends, and even more so, to me, Valery had found a way to rope me into the expedition, after all. There was simply no graceful way to get out of it.

Initially, at least, everything went like clockwork. Our travel through Moscow and out to Chukotka and the town of Anadyr went smoothly. Valery (shown in the color insert, image 12) was already there waiting for us. He welcomed us to the region with a great Russian feast that included reindeer meat and a delicious, salmonlike fish called Taranets char, which is native to the region and the cold Arctic waters.

The next morning, Valery took us to meet the local guides who would be taking care of us during the expedition—our driver, Viktor Komelkov and his wife, Olya Komelkova, who would serve as camp cook; and our second driver, Bogdan Makovskii. The two drivers seemed to come from opposite ends of the gene pool. Viktor was short, grizzled, and wiry and Bogdan was a head taller, clean-shaven, and built like a linebacker. None of them spoke English, but, through translation, I learned that all three were friendly, highly experienced, and committed to making the expedition a success.

The drivers took us to a large shed where the two massive vehicles were waiting. I stood frozen in shock when they swung open the doors. Valery's so-called trucks were actually colossal behemoths. One was blue and white, the other Princeton orange. The passenger section of the blue and white vehicle looked like the top of a large van while the bottom half looked like a massive army tank with a gigantic set of treads. The orange one seemed newer and more sleekly designed. They both looked bizarre, intimidating, and indestructible.

Marina translated as Viktor told us that each cabin could carry a driver plus six or seven other people. At top speed, he explained, they could reach up to fifteen kilometers, or nine miles, per hour. *It is going to take us forever to get to the dig site at that rate*, I thought.

The next morning, when our international team of seven Russians, one Italian, and five Americans gathered together to start the expedition, we discovered there was going to be a last-minute addition. Viktor and Olya introduced us to their beautiful cat, a Russian Blue named Bucks (see color insert, image 10). His job, they joked, would be to stand guard over the campsite. We instantly adopted Bucks as our mascot and he joined us for a photo in front of the strange-looking vehicles that were about to take us into the tundra (color image 8).

As soon as the photo was taken, we set off on the journey that was, for me, nearly thirty years in the making. Half of the team, including me, would be riding with Viktor in the blue behemoth and the other half, including my son Will, would be riding with Bogdan in the orange. Before we set off, Viktor tested his walkie-talkie to make sure he could reach Bogdan. Once two-way communication was established, everyone boarded the vehicles and we headed off for our adventure like a couple of slow-moving elephants.

When we first set off, Olya and her cat Bucks sat inside the cabin behind Viktor and me. But before too long, Viktor stopped to let them out. Olya put on some outer gear, placed Bucks in his cage, and, although she was shorter than all of us, adroitly climbed her way up to the top of the blue behemoth with Bucks in tow. From that point on, they rode outside together on a special seat designed specifically for that purpose. I was told that the two of them, both independent spirits, preferred to ride in the open air whenever Viktor drove across the tundra.

A short time later, both drivers suddenly came to a stop without warning. *Is something wrong?* I wondered, as Viktor ordered everyone out of both vehicles. I walked over to Will to see if he knew what was going on. He shook his head and shrugged. We watched as Olya and Bucks climbed down from their perch. Then Olya quickly got to work setting up a table and filling it with food.

You're kidding, I thought. *Is it already time for lunch? We barely got started.* Marina came over to explain. When embarking on a trip across the tundra, she said, it was a Russian tradition to stop and celebrate after crossing the first stream.

With a big smile, Olya then motioned everyone to the table where she had set out plates loaded with cold Russian pancakes, called blini, packed with meat and cheese. Along with vodka. There was to be a *lot* of vodka on the trip. During the celebration, we were instructed to pour out a little bit of vodka as an offering for the gods to ensure good luck. The rest of it was to be consumed. Since I drink only sparingly and figured that some of the others may not have been generous enough in their offerings, I made up for any possible shortcomings by discreetly contributing all of my share to the gods.

I soon learned that the tundra consisted entirely of mushy growth, heather, peat, and bog covering rigid tufts. It was surprisingly difficult

to traverse. Trying to walk just a few feet could be challenging. Boots got stuck and one could easily twist an ankle stepping on a hidden tuft.

I stared out the window as we drove along and realized that what looked like barren tundra was actually teeming with life. I could see an abundance of a white lichen growth called yagi, also called reindeer moss, which is a main source of food for reindeer. The animals have an enzyme in their stomach, I was told, that turns the mossy substance into glucose. There were relatively few birds, mostly quails and a few gulls. We also spotted some fast-running hares and a small polar fox wearing its summer coat of gray fur.

There were frequent streams and ponds along with deep, mud-filled tracks made by other vehicles, which looked like crisscrossing scars on the permafrost. Our trucks had to try to move around the earlier tracks for fear of getting stuck in one of the muddy ruts.

I was surprised to see that there were a lot of flowers, the most common of which was a delicate snow-white flower called Arctic bell-heather. Their silky white petals formed a bell-like shape that sat atop thin stems that gave way gently to the wind. It was like gazing at a beautiful sea of white oscillating back and forth. I nicknamed them the "laughing flowers of the tundra" because their swinging and swaying movements made it seem as if they were giggling uncontrollably at the silly humans attempting to venture through their terrain.

Kilometer after kilometer, Viktor and Bogdan battled up and down the tundra's uneven tufts at about four miles an hour. After a few hours, the orange vehicle's engine started sputtering and stalling. Viktor had to repeatedly stop our blue truck as we waited for Bogdan to catch up. The two of them decided that a previous driver must have filled up the orange truck with the wrong grade of diesel fuel. They tried to compensate by syphoning in fresh oil from another drum, but nothing seemed to work. The frustrating stop-and-go sequence went on and on until nearly midnight, and by that time Bogdan had fallen so far behind that we could no longer see him. Even worse, the walkie-talkies had inexplicably stopped working. So we had managed to lose all sight and sound of half our team.

We had been on the road since 6 a.m., and it was now approaching midnight. So Viktor finally decided to call it a day. He drove the truck down a steep incline and stopped near the edge of a small river. It was then that I jumped off the truck, only to be enveloped in a suffocating cloud of mosquitoes.

Bone-tired, futilely fighting off my tiny attackers and gasping for breath, I was out of my element and losing my grip. In a searing moment of utter panic, it suddenly occurred to me that my son and the rest of the team that I was supposed to be responsible for was still missing.

Frustration, exhaustion, and panic, laced with a sense of absolute dread. An avalanche of emotions swept over me, unlike anything I had ever experienced before.

And where was my son?

EIGHTEEN

FOUND

Increasingly frantic, I swung my arms wildly in all directions to swat away the mosquitoes that were relentlessly attacking me. I stood on the muddy hillside, craning my neck for any sign of the orange behemoth. The wind howled like a monster in my ears.

"Paul."

It sounded like someone was calling my name, but I brushed it aside. Panic and fear were making me imagine things.

"Paul. *Paul.*" No, that voice was real. It sounded vaguely like someone with a Russian accent, but it was hard to tell because of the wind. *NOT NOW*, I thought. *GO AWAY.* I did not want anyone to see me in this nearly hysterical state. I had completely lost control of my emotions.

But the Russian voice became more insistent.

"*PAUL!*" Now he was right behind me, demanding my attention.

Exasperated, I spun around and was startled to see Valery Kryachko standing within arm's reach. He did not speak any English and knew that I did not speak Russian. So without saying a word, he stepped forward and plopped a camouflage hat on my head.

I must have looked confused. Valery studied me for a moment, and seemed to realize that I did not understand what was going on. So he reached up and began fiddling with the brim of the hat. I stood there silently, having no idea what I was supposed to do, until I finally noticed that Valery was pulling black netting from the top of the hat down over my face. He tightened the cord along the bottom of the netting until it closed around the bottom of my neck, which instantly created a barrier between me and my attackers. He took a step back to look at me and nodded approvingly. Then, without saying another word, Valery turned and started making his way back down the hill.

For the first time since getting out of the blue behemoth, I could finally breathe normally. I could stop waving my arms all around like some kind of crazed human windmill.

And then, things got much, much better. I suddenly heard the low growl of a diesel engine approaching from the distance. I swung around to face the hillside and watched as the orange behemoth suddenly flew over the top of the hill, careened past me and shot down to the riverbank to join its blue partner. My anxiety was replaced by euphoria the moment I saw it go by. My son was safe, my team was safe. And thanks to Valery, I had not been asphyxiated.

I followed the orange truck down the muddy hillside as quickly as I could. Will, being an experienced field geologist, emerged fully outfitted in his mosquito gear. When I asked about the whereabouts of my own gear, which I had mistakenly left with him when we had set out that morning from Anadyr, he cheerfully informed me that he had put it to very good use cushioning his computer and camera equipment.

A few moments earlier, when I had been suffocating in a cloud of mosquitoes and terrified about what had happened to him, I might have had trouble fully appreciating the joke. But now that he was safe and the mosquitoes had lost their hold over me, we were both able to laugh at the absurd situation. Valery's hat was to become a regular fixture on my head, and would keep me sane and calm for the rest of the journey.

Dinner the first night out was a hot soup of ramen noodles and warm reindeer ribs served along with more meat wrapped in blini. Everyone ate rapidly. After having spent the first sixteen hours of the expedition bouncing along the tundra at a sluggish four miles an hour, everyone was eager to get some sleep.

Except for me. Once our sleeping bags were laid out in the back of the blue truck, I spent part of the night lying flat on my back, shoulder to shoulder between Will and Vadim. I managed to last about three hours in that position, but could never really get to sleep because the vehicle was tipped at such an angle that blood was rushing to my head. *Would every night be this bad?* I wondered.

I finally concluded sleep was hopeless. Taking Valery's mosquito hat with me, I quietly slipped out the back of the truck and climbed up to the perch where Olya and Bucks had spent most of the first day. I stared across the strange, moonlit terrain and pulled out my logbook to record my thoughts about our first day on the tundra:

> *The snow-cats, or at least I am calling them that, seemed impressive*
> *when we first saw them yesterday, but today, against the huge expanse*
> *of tundra, they seemed completely insignificant. The ride was one*
> *giant amusement park ride lasting all day . . .*

An hour or so later, Sasha woke up and joined me atop the blue behemoth. I had originally recruited him solely as a Russian translator, but Sasha had other qualities that would make him a real asset to the team. Tall and athletic, he had a wealth of outdoor experience that would enable him to play various roles in the expedition. With his wavy blond hair, ever-present smile, and buoyant enthusiasm, he would help keep everyone's spirits up and ensure smooth communication with our Russian colleagues.

Sasha also possessed a basic understanding of physics from his undergraduate days and was insatiably curious. As we sat together atop the vehicle, he took the opportunity to ask a lot of questions about how I had first conceived of quasicrystals, which had been such a radical idea at the time, and why I thought it was important to now search for a natural sample.

Soon after we got back on the road the next morning, we began experiencing more frustrating mechanical problems. The orange truck had suffered most of the problems the first day because of issues with diesel fuel. On the second day, it was the blue vehicle's turn. One of the huge tank treads separated from the wheel sprockets, much the way a chain might detach from a bicycle gear. A bicycle could be easily turned upside down in order to reattach the chain. *But this monster?*

Bogdan was prepared for such emergencies, and a thick log suddenly materialized from underneath his truck. He took out an ax and chopped off a sizable piece of the log, which he then split into quarters. He put the split pieces of wood into the gaping spaces between the gear teeth and the detached tread.

The team was transfixed as Viktor climbed into the blue truck and slowly drove forward, drawing the relatively small pieces of wood around by the action of the gear. The wood traveled nearly full circle until *SNAP!* The tread suddenly reconnected with the gear teeth and crushed the wood to pieces.

The most frightening aspect of the process was watching Bogdan reach his bare hands into the moving machinery. He had to keep jamming the wood to keep it in place as Viktor drove forward. There were several heart-stopping moments when I thought he would lose a few fingers if not his whole hand. But for the two drivers, the operation was as routine as fixing a flat tire. They would repeat the process several times during the drive, and it was always terrifying to watch.

By the afternoon of the second day we could finally see the Koryak Mountains, home to the Listvenitovyi Stream, in the distance. That was our ultimate destination. But it was clear that we would not even come close to reaching the outskirts of the mountain range by the end of the day, given the slow pace at which we were lumbering across the tundra.

By midday, we came across a natural gas facility that had been built to support a Russian mining operation. We were hoping to be able to clean up and enjoy lunch at the company cafeteria. But no such luck. Fresh food was in short supply at the remote location and, according to the manager, all of their spare food was frozen to help it last longer. We were turned away. But the manager promised to make up for it on our return trip, provided we remembered to call ahead to let them know we were coming.

I couldn't help but smile. Even in the middle of nowhere, it seemed, one needed to make a reservation.

We continued driving for several more kilometers and came across a defunct drilling station, seen on the facing page, that appeared to be the complete opposite of the modern one we had just left behind. The dilapidated setting reminded Will and me of something out of the apocalyptic *Mad Max* movies with rusted oil derricks and remnants of old vehicles and oil cans strewn everywhere. But the site only looked deserted. It was still being used as a refueling station, and Viktor had arranged to pick up several drums of diesel fuel to replenish our supplies.

Once we collected the oil drums and headed back on our way, Viktor and Bogdan ran into one dead end after the other. First

a crevasse, then a quarry, or multiple times an impenetrably thick growth of vegetation. Every time we hit an obstacle we had to turn around and retrace our way back to the beginning before heading off again in another direction.

We continued driving back and forth for several hours as if trapped in a maze until it finally became obvious that it was becoming quite late and we were getting nowhere. By now, Viktor and Bogdan were mentally spent and physically exhausted, so we headed back to the beaten-up headquarters at the *Mad Max* station for the night. Our second day of driving was over. But we were already a full day behind schedule, thanks to all of our mechanical and directional problems.

Olya hastily pulled together a dinner that we ate inside one of the trailers in *Mad Max-ville*, when Viktor hurried back inside to tell us about yet another problem. He had been checking the two new fuel barrels he had stored in the back of his truck and discovered they were both defective and leaking oil.

If we had continued driving toward the Koryaks, we would have been heading for disaster because the barrels would have continued leaking and gone unnoticed. We could have lost our entire fuel supply and wound up stranded in the wilderness. Worse, a leaky oil drum could have set off an explosion in the truck. So the frustrating delay turned out to be a blessing in disguise.

"Maybe I deserve credit for our good luck," I told Will jokingly. "Pouring out so much of my vodka as an offering to the gods during the stream crossing ritual might already be paying off!" Will just rolled his eyes, an expression every parent can relate to even if they are not camping in the middle of the tundra.

With the leaky oil drums safely replaced, we set off the next morning and quickly found a direct route to the Koryak foothills. The soggy tundra under our treads was now replaced with solid dirt and rock and we began to fly, if you can call fifteen kilometers per hour flying.

As we continued to climb into the foothills, we had our first encounter with a couple of the four-legged natives. We spotted two Kamchatka brown bears studying us from a distance and could see that it was a mother and cub. Even from a distance, it was clear they were both gigantic. Our diesel engines were so loud that the bears could hear us from across the valley and the cub was so curious that it stood up on its hind legs to get a better look. I was grateful to be in the truck and not out in the open because female Kamchatka bears, like every other bear species, are known to be fiercely protective of their offspring.

The mother eventually coaxed her cub into leaving, and as they ran off together I marveled at their speed and power. Kamchatka brown bears can run up to thirty-five miles an hour, much faster than our vehicles. The mother and cub were able to keep up their strong pace until they were far from sight.

If we were ever unlucky enough to encounter one of the Kamchatka bears at close range, I thought, *there would be no point trying to outrun it.*

Chris Andronicos, one of the most experienced outdoorsmen on our team, had worked as a field geologist in a lot of regions with grizzly bears. So before we left Anadyr, I had asked him to give the rest of the team "bear safety lessons." Chris kept it simple:

Lesson 1: When encountering a Kamchatka brown bear at close range, it really makes little difference what you do. You are probably dead.

Lesson 2: Try to avoid being at close range to a Kamchatka brown bear. A good way to accomplish Lesson 2 was covered in the third lesson.

Lesson 3: Keep in groups of three or more and make a lot of noise wherever you go. Bears have bad eyesight and perceive a group of people moving together and making noise as a single beast that is much larger than they are. They will tend to move away, assuming there is another promising food supply.

The scenery changed dramatically as we traveled farther into the Koryak range, where the slopes were brown and almost entirely barren (pictured on the next page). Chris explained that it was mantle rock that was rich in olivine and peridotite. Similar to certain mountainous regions of California, he said, the soil was toxic because of the high concentration of nickel. Trees and thick vegetation could not survive. The exotic, otherworldly setting almost made me wonder if Chris's preferred theory that the Florence sample was made on Earth might be true after all.

As we traveled deeper into the foothills, we came across a series of stunning river crossings (see the color insert, image 9). The riverbanks had dramatic drops of twenty feet or more, which was much steeper than anything we had experienced on the drive so far. The behemoths

could manage the precipitous drops, but Viktor began cautioning us to hold on tight as we toppled over the edges.

I had been sitting in the front bench seat next to Viktor throughout the past few days, spellbound by his driving technique and trying to anticipate whether he would swerve this way or that to avoid an upcoming rut or some other obstacle. By now, I had become familiar with all of his defensive driving techniques. So I was alarmed when he suddenly shifted into high gear after we completed one of the steep drops over a riverbank.

Standing before us was a forest of tall trees. I gasped as Viktor floored the gas pedal and aimed straight for them at our top speed of fifteen kilometers.

"Viktor, where is the trail?" I shouted, as we surged forward. My tone of voice must have said it all because Viktor did not wait for the translation. He gave me a wry look as if to say, "Trail? Who needs a trail?" And with that, he drove directly into the forest.

The trees fell over one by one like flimsy pieces of cardboard.

When I turned around to inspect the damage, I discovered that the trees I had thought impenetrable were actually quite flexible. They sprung back up behind us as if closing a trapdoor. Evidently, the trees were young enough and their trunks elastic enough so there was no permanent damage.

Next, we made a brief stop in a strange-looking valley to pick some wild mushrooms for Olya's kitchen. Having spent the last few hours cooped up inside the trucks, the team enjoyed a chance to stretch their legs in the field of freakishly large mushrooms, some of them a whopping ten inches in diameter. Marina, pictured above, and Olya led the effort.

Chris and I stayed behind to study the maps and current GPS positioning. We concluded that we were not making good enough progress. So I made the extremely unpopular decision to cut the mushroom-gathering adventure short so that we could get back on the nonexistent "road." And once we did, we came up upon the mighty Khatyrka River, the deepest and widest river in the Koryak mountain range.

I'll never forget that moment. As we approached the river, the normally quiet Valery Kryachko leaned forward from the backseat and said something to me in Russian.

"Valery says there is considerable uncertainty whether we can cross the Khatyrka," Sasha translated. "The depth of the water varies from year to year and from season to season."

What? Is he serious?

I thought back to all the planning meetings over the past six months. We had discussed government approvals, food and fuel supplies, weather conditions, and bears. But during all those meetings, we had never discussed the terrain we would be dealing with and the challenges it might present. I had been told there would be roads. But there had been no roads and now a pretty wide river was blocking our path.

Why am I only hearing about this now—just as we are approaching the water's edge? I thought with disbelief.

Sasha continued to translate, this time for Viktor.

"The vehicles were designed to float," he said. "Except for the fact that. . . . well, except for the fact that no one has ever tried to float them across a river as wide as the Khatyrka when they were carrying so many people and so much weight."

It was a nice time to let me know, I thought. I had no notion until that exact moment that we might need to float across a river hundreds of meters wide and of indeterminate depth to get to our destination. I had been concerned about making it to camp by midnight. Now, I was worried the Khatyrka would keep us from ever getting there at all.

How in the world did we get this far without having had this discussion? I thought, shaking my head.

We stopped on the bank of the Khatyrka and disembarked while Viktor and Bogdan set off in the orange vehicle to scout the way forward. In the interim, the rest of us had the opportunity to share a late lunch. The team's spirits were buoyed by the delicious smell of the freshly picked mushrooms Olya cooked as part of the meal. So I kept my concerns about the river crossing to myself and let the team enjoy their lunch, most of them unaware of the peril that lay ahead. Everyone would find out soon enough, I figured.

When the two drivers returned, they announced that they had identified what might be the shallowest route across the Khatyrka. So we packed up our gear, climbed back into the two behemoths, and braced ourselves for the weighty experiment.

I had no idea what to expect as we surged into the river waters. It was a strange sensation to feel the strength of the river as it periodically took control of our mammoth vehicle as if it were a toy floating in a bathtub. The current lifted us up and off the riverbed, and then carried us sideways downstream before letting us back down. For the next ten minutes, both of the behemoths drove, then drifted, then drove, then drifted their way across the river.

Eventually, both vehicles made it to the other side. I breathed a deep sigh of relief as we drove out of the water and up the riverbank.

What was the next surprise? I wondered. *What else did Valery forget to tell me?*

We continued on, with the blue truck once again racing ahead of the orange. Years earlier, Viktor had taken part in a mining expedition in the area and was familiar with the path. So he knew that we were getting very close to our intended campsite when he spied an odd row

of poplar trees. He turned to follow the row of poplars until we suddenly reached a clearing and riverbank.

Here was the Iomrautvaam River, a tributary of the Khatyrka River, along which we would be camping. Viktor barreled across the waters, drove onto the riverbed, and hit the brakes.

I climbed out of the truck and looked at my watch: 8 p.m. Four days of driving for sixteen hours at a time, some of it frustrating, all of it uncomfortable, and we had finally arrived at our destination. Everyone cheered and offered Viktor congratulations.

We didn't waste a moment. Without waiting for our companions in the orange truck to arrive, we left Viktor behind to take care of the camp and headed off for a short hike. With Valery leading the way, we walked along the edge of the Iomrautvaam until we reached the place where a small stream, the narrow Listvenitovyi, flowed down into it. The growth was too dense for us to turn upstream and walk along the Listvenitovyi itself. But Valery told us we would be using a different path the next day that would take us to the dig site.

I was elated to have finally arrived, recalling the many hours spent staring at maps of the region with Lincoln Hollister and dreaming of how we might be able to organize a team of experts to explore it. Now it was actually happening and I was leading that team. Despite what Valery told us, I had the urge to march right through the dense growth blocking our path and start digging right away. I felt like a child being told they had to wait to open their birthday present.

As we strolled back to our new campsite, Valery made sure all of us got a good look at the huge, fourteen-inch bear tracks along the trail. Kamchatka bears were nearby, drawn to the river's fish supply. It was fair warning that we would need to take every possible precaution to avoid meeting them.

By the time we returned from the Listvenitovyi, the rest of our expedition team had arrived. In a remarkably short amount of time, our gear was unloaded and the first tents erected, including a dinner tent draped in mosquito netting.

Most of the team members were assigned tents with just enough room to sleep one or two people. But Valery had thoughtfully selected a much larger tent for Will and me that was almost tall enough for the two of us to stand up in. It had double flap doors and was equipped with its own heater and, thankfully, its own bug killer. Valery even ran electrical power to the tent from a nearby generator so that we could recharge our computers. He knew that with me, he was dealing with a complete indoorsman and seemed to want to do everything possible to make my first camping experience a pleasant one. I was grateful that he took such good care of me. I was also grateful to my son for accepting the embarrassingly luxurious accommodations, even though he was an experienced camper accustomed to rougher conditions.

The campsite itself was a few hundred feet from the Iomrautvaam River, close enough that we could hear its waters flowing while inside our tent. In the distance, we were flanked by the Koryak Mountains, the source of the Listvenitovyi Stream. The area where our tents had been set up was fairly nondescript and ordinary looking. It was flat, with sprouts of shrubs popping up here and there. Across the river, the shrubs and growth were somewhat taller, a few as high as twenty feet. The place felt strange and unfamiliar to me, but not threatening.

Will and I spent the first night organizing our tent and discussing our plans for the next day. In contrast to the cramped conditions in the back of the truck the last four days, there was more than enough room in the tent for our sleeping bags and I had no trouble falling asleep.

The next morning, our expedition began in earnest. We trudged off for the dig site, where we would try to replicate Valery's 1979 success. I made sure to put Valery's mosquito hat on before going outside because I never wanted to repeat my nearly suffocating introduction to the tundra. Will and I had brought along enough DEET to supply an army, but the lotion seemed to attract mosquitoes as if we had coated ourselves with delicious candy. All of our Russian teammates, as well as Chris Andronicos, were used to such environments and seldom resorted to using netting or bug repellent.

After a fifty-minute hike, during which we made lots of noise in order to announce our presence and keep the bears away, we arrived at the stream (pictured in color image 13). Almost immediately, Valery pointed out the legendary blue-green clay described in the Razin paper. It was astonishing. He casually reached into the mud to scoop out a handful of the clay, then rolled it into a ball and passed it around the group. Once it was compacted, the clay seemed to have the same texture as chewing gum or Silly Putty.

Holding the ball of blue-green clay in my hands for the first time was an indescribable thrill. I can only imagine that Luca felt the same way. It was tangible evidence that we were now very close to the source of the natural quasicrystal that had figured so prominently in our lives and our minds for the last two years.

I was still lost in thought when Valery interrupted. "We're not there yet," Sasha translated. It was the same blue-green clay, but not exactly the same spot. So for most of the next hour, we followed Valery as he zigzagged up and down the stream, walking a half kilometer this way and that.

There was a specific landmark Valery was searching for that he had not mentioned until now, a spire of rock about fifty feet tall. It took quite a bit of searching, but once he spotted it, he was visibly relieved and moved decisively toward it.

About a thousand feet from the spire, Valery stopped along the bank of the Listvenitovyi and declared that this was the place where he had found the Florence sample.

After thirty-two years, and in defiance of all the naysayers I had consulted, Valery had indeed managed to bring us back to his original dig site. It was a moment of intense personal pride for him and one that I will never forget. Luca and I took a photo together holding our hands raised high in victory (color image 14), the image from my dream long before we ever knew of Valery's existence or that we would be taking this journey together. To say that the first morning went better than I ever could have expected would be an understatement.

We named Valery's spot the "Primary," meaning it would be the primary location where we would begin digging and collecting samples, hoping that lightning might strike twice and we would discover another natural quasicrystal.

We hiked back to camp and discovered that there was an unexpected benefit of camping next to the Iomrautvaam River. While we were busy searching for the dig site with Valery, our drivers Viktor and Bogdan had set up hundred-foot-long fishing nets across the river. It was the summer spawning season, so it only took them a few hours to catch enough salmon to feed the entire team. The abundance of salmon helped confirm our suspicions about the presence of all the hungry bears in the neighborhood.

At dinner that night, Olya was able to serve freshly caught salmon and fresh caviar. It was the orange-colored caviar that is so common in Kamchatka. Everyone was impressed, especially Chris, who declared that this trip was already setting a new standard for geological expeditions. I had never been much of a caviar fan, but was instantly converted.

Caviar, especially salmon caviar, spoils quickly, and is typically treated with salt and preservatives to prolong its shelf life. Up until now, all I had ever tasted was that overly salted version. But there is simply no comparison between the preserved caviar one buys in a can and what we were treated to while dining alongside the Iomrautvaam River.

As I looked around the table, abundant with delicious food, I began to truly appreciate all of the contributions Olya was making to our effort. She had the ability to be warm and welcoming but at the same time, could be very business-like. As a lawyer, Olya had worked with our Russian colleagues to expedite the complicated permit process with governmental authorities. As an organizer, she had helped secure our transportation and coordinate our stay in Anadyr. Now as a cook, she was making another tangible contribution by helping energize and inspire everyone.

I divided the group into three teams: the diggers and the panners, who would both work along the stream, and the mappers, who would explore the area and map the local geology. The only one who wound up without a specific assignment was me. I gave myself the job of "floater" so I could move from team to team to help solve problems and ensure we were working together as effectively as possible (see color images 15 to 18).

The next day, I decided to spend part of the afternoon with Glenn, Chris, and Mike on the mapping team. They began working near the Primary trench and stormed their way up the Listvenitovyi Stream as fast as possible, taking note of rocky outcrops along the way. Then they slowly worked their way back downstream, systematically recording the locations and geological properties of those rocky outcrops. After a few hours moving down the stream in that methodical way, we could see the fire and steam rising up in the distance that was being created by the panning team.

Once we arrived back at the Primary, it was obvious that Will, who was doing the digging, was having a great time working with Luca and Valery, who were doing the panning. Will would dig up enough blue-green clay from the riverbank to fill up a large pot. Then Luca would add water from the stream and bring the mixture to a boil over a hot fire. The boiling water made the clay less sticky, so one could reach into the water and break up the clay until it became more like a thick sand.

Valery, the master gold panner, would take the pot of thick sand to the Listvenitovyi where he would swish the huge pot in the cold mountain water. To the uninitiated, the swishing operation gave the appearance that Valery was washing everything out of the pot. But in reality, he was carefully separating out the densest grains because he knew the material we were looking for was denser than most terrestrial materials. Once he was satisfied, Valery would pour the remaining contents into a broad, V-shaped wooden dish and repeat the swishing process in the stream in order to further separate the denser particles. Every now and then, Valery would bring the wooden dish close to his face to look for any unusual or shiny bits. After another repeated round of swishing, he usually wound up with a palm-sized amount of material, which he would pour into a small bowl before repeating the process. Once Valery was satisfied, he would pour the small bowl into a small metal cup and after a final round of swishing, the meager amount of remaining material was given to Luca, who would boil away the water. In the end, the final product was a dry

powder of promising mineral separates.

Ultimately, Luca would pour the dried contents into a plastic bag and assign it a number along with the date and source location. The labor-intensive process would be repeated over and over and over again until the very last day of the expedition to obtain as much sample material as possible for later testing.

In addition to our mapping, digging, and panning

teams, Marina, Vadim, and Sasha formed an independent team to identify additional locations of blue-green clay up and down the stream. They also brought samples to Luca and Valery for boiling and panning.

After checking out all the activity, I went back to my tent and began adding notes to my logbook. At the end of the day, Will, Luca, and Valery returned from the stream. Will threw open the tent flap and asked me to hand him the video camera. I noticed that he had a funny grin on his face, but all he was willing to say was that I should come outside to watch him film Luca. It seemed like an odd request, but I played along. Luca was standing in front of the dining tent where Glenn, Vadim, Chris, and Mike were chatting and having tea. I noticed that Luca, just like Will, also had a funny grin on his face.

After setting up the camera and hitting the record button, Will called out to Luca so that everyone could hear:

"Luca, what happened today?"

"I have found a grain in one of the clays," Luca said, "that has shiny metal connected to silicate. I think it is a quasicrystal candidate."

Caught by surprise, I was speechless. *Was Luca serious? Was it possible that we had been successful on our very first day in the field?* I did not know what to think, but rushed forward to give him a big hug.

Luca told me it had been a group effort. He, Will, and Valery had been focused on their digging and panning work at the Primary while Marina and Sasha were searching for other promising dig sites along the stream bed. After a while, Marina had come back to the Primary to tell them about a spot on the other side of the stream, which was later dubbed the "Green Clay Wall." Will immediately stopped what he was doing and walked across the stream to where Sasha was waiting. Together, the two of them dug out a bucketful of greenish clay, which they brought to Luca and Valery for processing.

Something caught Luca's eye as he took a quick look through the panned grains. He spotted a peculiar grain that was composed of small metallic bits attached to blackish minerals. Luca immediately showed it to Valery and Will, who also noticed the shiny metal.

It seemed auspicious, but everyone knew that there was no way for us to positively determine while still in the field if Luca had discovered a meteorite, much less a tiny quasicrystal within that meteorite. That would require the kind of advanced microscopes we had back home in our laboratories. But I also knew from experience that Luca had a great eye. *So maybe, just maybe, he is right*, I thought.

After dinner, we took out the crude geological microscope that Valery had brought with him and tried to examine Luca's grain. It looked promising to me, but Chris and Glenn were skeptical. They were certain that it was chromite, a common terrestrial mineral. A friendly debate ensued for about an hour, but the truth would have to wait until we returned home and could study the sample properly. Even though we could not be sure of the outcome, the potential discovery helped inspire the team for the rest of the trip.

At last, everyone went to bed for the night except Will and me. Will was busy organizing the photos and videos he had taken, and I was too energized by all the events of our first day in the field to relax. I stepped outside the tent for some fresh air and walked over to the river's edge where I had a clear view for miles around. I watched as a low-lying fog crept its way across the river valley. A starkly beautiful quarter moon rose above the fog in the gorgeous, crystal-clear, midnight blue sky. It was nature at its most exquisite, and I called Will outside to take a photo (shown in color image 19).

The two of us stood together on the riverbank for a long time. This exotic place was unlike anything either of us had ever seen before and we were both mesmerized by the hauntingly beautiful view.

NINETEEN

NINETY-NINE PERCENT

By morning, the distant fog Will and I had admired the night before had turned into a dreary rain. As temperatures plunged to near freezing, the team took turns at breakfast huddling around the kitchen stove for warmth.

The bad weather put our digging and panning efforts on hold. But I watched with quiet concern as the mapping team geared up for their first long-distance hike to one of the nearby peaks.

Chris was in charge of the team. As a structural geologist, he planned to locate and identify rocks that could be used to determine if there was any evidence of a superplume or unusual geological phenomena that could provide a terrestrial explanation for our sample. Alternatively, rocks would also be examined for micro-cracks that might indicate the impact of a large meteor. Thick vegetation at the base of the mountains covered most of the rocks except for a few outcrops. So Chris had decided to take Glenn and Mike up to a nearby peak where some bare rocks were more accessible.

It was a huge undertaking so I could understand why the mapping team did not want to be sidelined by bad weather. But I was worried about the decision to charge ahead because we would never be able to help if they ran into serious trouble. We were prepared with basic first-aid supplies, of course. But we could not cope with any critical injuries. The emergency plan was to use our satellite phone to call for help. But all of the worst-case scenarios kept running through my mind.

What if the storm worsened? What if we could not get a satellite signal? What if the weather was so bad that the rescue helicopter was grounded and emergency teams could not arrive?

As I feared, the storm worsened considerably by the time the mapping team made it to the top of the nearest peak. They were forced to turn back. And their return to camp would be especially grueling. The three of them had to maintain their footing as they carefully worked their way down from the peak in a freezing downpour. Once they reached the bottom, they had to slog their way through acres of mud and muck while fighting against the thick wet brush, which was waist-high at points and teeming with mosquitoes.

They finally made it back to camp without injury but were soaked to the bone and clearly exhausted. Once they changed back into dry clothes, Chris and Mike appeared none the worse for wear. But the same could not be said for Glenn, who was accustomed to working in an indoor laboratory. He had done virtually no geological field work since joining the Natural History Museum staff twenty-seven years earlier. Glenn was also the oldest of the group and not particularly physically fit. Within half an hour, he was shaking uncontrollably.

Marina and Olya immediately recognized the signs of hypothermia and swung into action. They sat him down, covered him with blankets, and began administering hot tea and soup, along with vodka, the Russian cure-all. *Passive rewarming*, I thought. *This is all we can do.* Glenn no doubt sensed he was in danger because he seemed to panic and began snapping at the women whenever they tried to help. *Irritability—a symptom*, I thought. *So is resisting help, which is going to make everything more difficult.*

The team gathered around and watched as the frightening scene unfolded. I, for one, felt helpless. *What if this doesn't work?* Fortunately for Glenn, despite all of his complaints, our two Russian colleagues gently persisted in keeping him as still as possible. It was a long fifteen minutes before the shaking finally began to subside and the color slowly began to return to his face. Marina and Olya's swift intervention prevented Glenn's hypothermia from progressing beyond the mildest stage, but he would need quite a bit of rest before he could fully recover.

Thank goodness he was back in camp when this happened and not in the middle of a hike, I thought with relief.

Disaster averted, the team dispersed and I walked back to my tent to record the day's events in my logbook. Glenn would be all right, I concluded. But I would need to rethink his role in the expedition.

Blam! My note-taking was suddenly interrupted by the sound of a loud blast. *What now?* That sounded dangerously close to my tent.

Blam! After the second blast, I knew it was gunfire and my thoughts turned to the only weapon in camp, a modified AK-47 that our Russian drivers kept on hand as a precaution against a possible bear attack. The Kamchatka brown bear's fur is apparently so thick that it can blunt the effect of a smaller weapon.

Kamchatka bears are enormous, fearsome creatures. The largest males can weigh up to 1,500 pounds and are absolutely gigantic. They stand ten feet tall on their hind legs, a posture they often adopt to make full use of their keen sense of smell. During the summer and fall months, the bears are instinctively driven to consume an enormous number of calories in order to store enough fat to survive the winter hibernation.

Three years before our expedition there had been a deadly bear incident at a large compound south of our location. Salmon poachers had severely depleted the fish supply in the area that summer, and thirty starving bears made a concerted attack against a mining camp in search of food. Hundreds of geologists and miners scattered in panic. The bears easily outran two of the men, who were gruesomely killed and eaten.

We had been finding fresh bear tracks all around the campsite and knew there were groups of bears nearby, gorging on fish in the Iomrautvaam River. So it would not have been a complete surprise to me if the shots I was hearing were actually meant for a bear.

I rushed outside and saw that everyone, including Will, was crowded together about fifty feet behind my tent. No one was running for cover and everyone looked relaxed, so I quickly surmised there was no immediate threat. When I asked Will what was going on, he told me that Viktor and Bogdan were holding target practice. The drivers usually kept the Kalashnikov loaded with large cartridges for protection against bears. Now they were using smaller caliber bullets and had set up empty vodka bottles along the riverbank as targets.

I watched carefully as other team members took their turns with the rifle. Viktor wanted everyone to try hitting the targets, including me. I did my best to refuse, explaining that I had never fired a gun of any kind much less something as powerful as a Kalashnikov.

"No," Viktor insisted with a smile. "No exceptions." Everyone would have to take three shots.

He handed me the rifle and I looked at it warily. Normally, I would want nothing to do with such a weapon. But under the circumstances, I felt there was no choice. So I brought the Kalashnikov up to my shoulder and took aim. Having never fired a weapon before, I made the rookie mistake of rotating the barrel toward my face to make it easier to look along the sight. As a result, and to the great merriment of the rest of the team, neither of my first two shots landed anywhere close to the targets. Everyone roared with laughter.

Viktor reloaded the rifle for my third and final shot and quietly murmured something in Russian to Sasha. "You should adjust your grip," Sasha translated. "You need to aim with the gun sight straight up, looking straight over the top of the rifle."

I nodded wordlessly, knowing the advice would be of no help. The problem had nothing to do with how I was holding the rifle. The problem was that my eyes were not sharp enough to see any of the targets. But there was no point trying to explain. At this point, I just wanted to get the humiliating experience over with as quickly as possible.

I took aim in the general direction of the bottles and fired off my last shot. In response, the team erupted with even more laughter than before. I stepped back, and sheepishly gave up the rifle.

A few hours later, the subject of target practice came up during a conversation with Will in our tent. I told him I was sorry if I had caused him any embarrassment when I proved to be a total failure as a sharpshooter. He gave me a puzzled look and said, "What are you talking about? You hit the target on your last shot!"

I was flabbergasted. The team had not been laughing at me after my last shot. They were cheering for me.

That was impossible, I thought with a smile. If I had managed to hit the target, it was literally a case of blind luck. So blind, in fact, that I could not even see the vodka bottle shatter. But I was more than happy to accept bragging rights. I had successfully fired a Kalashnikov in Kamchatka, which is a claim few other theoretical physicists can match.

Glenn had spent the day resting and rebuilding his strength. I was glad to see that he was doing well enough to rejoin the group for dinner, even though he still seemed tired and worn out. As the evening progressed I noticed that Glenn, who was normally a strong if not combative personality, seemed vaguely dispirited. Sensing that he might be concerned about his future role on the mapping team, I decided to help ease his mind.

"No more long hikes," I told him firmly. "We have plenty of other important tasks for you to do closer to the camp."

Thankfully, it was one of the few moments in our professional relationship that Glenn did not try to argue with me about one of my decisions. Instead, he was relieved and readily agreed. The unspoken problem, of course, was that I had not yet figured out what his next assignment would be. First, I would need to make sure that he had fully recovered.

Under clear skies the next morning, we were able to get back on schedule (color images 20 to 23). Will, Sasha, and I joined forces digging a new trench at the Primary. When Valery was at the site in 1979, only the very edge of the riverbed had been bulldozed. In the years since, Russian gold miners had removed an additional ten or twenty yards of dirt from the side of the stream. That gave us a good head start, because it meant we had less soil to remove to reach the blue-green clay we knew was buried in the hillside. The clay was our target, because it was linked to Valery's original discovery.

Digging was tough going, because the blue-green clay was heavy and sticky. In less than an hour, all of our regular shovels were broken. From that point on, digging at the Listvenitovyi was done with a combination of flimsily repaired shovels, trowels, and our bare hands.

Valery had chosen to work at a different site just downstream from us. It was pristine and, unlike the Primary, had not been contaminated by any mining activity. We had all been working for about an hour when Valery suddenly started shouting to us excitedly in Russian. "He wants us to see what he has found," Sasha said.

We walked about fifty meters downstream and Valery showed us the hole he had dug next to the water. The three of us watched as he reached inside with both of his bare hands and pulled out a thick ball of mud. After a few moments, it began to solidify and Valery gave us a knowing look. We stared at his hands as he cracked the ball open

like an Easter egg, revealing the hidden prize inside. Blue-green clay! Valery grinned as we cheered his latest discovery, which immediately reset our agenda for the rest of the day.

Will and Sasha left the Primary and spent the next few hours working at Valery's new dig site. We planned to continue to excavate the area, so they built a thick wall of clay around the hole to keep the stream from refilling it with water. Will seemed especially determined to extract as much clay as humanly possible. The site, pictured on the next page, would eventually come to be known as "Will's Hole" in honor of his single-minded dedication.

Instead of taking a break and heading back to camp for lunch with the rest of the team, Will stayed on-site that afternoon and nibbled on snacks he had stashed in his backpack. Even though he was working alone at the stream, I knew he was not in any danger from the bears we assumed were roaming nearby. The bears had no interest

in the relatively puny waters of the Listvenitovyi when there was a rich supply of salmon swimming downstream in the Iomrautvaam River.

Eating outdoors was more complicated than Will expected, though, because it meant trying to cope with the hordes of mosquitoes attracted by his breath. Exasperated, Will finally tied a bandanna over the bottom half of his face. Even so, every time he tried to put food in his mouth, he wound up inhaling a mouthful of mosquito appetizers.

Later that afternoon, we encountered our first major scientific challenge. Glenn and Luca hiked out to the dig site to study some of the samples, and after much checking and double-checking, they came to a startling conclusion.

"It is entirely possible that we are on the wrong track," Glenn explained to me. "Blue-green clay might not be that relevant to finding more samples." Their finding was a major surprise and would have major repercussions on the rest of the expedition.

We had been speculating about the importance of the clay for the last two years, ever since the very beginning of our investigation. Luca and I had first learned of its existence in the scientific paper Leonid Razin and his coauthors published announcing the discovery of khatyrkite and cupalite.

How important was blue-green clay to the discovery of meteorite samples? we always wondered.

At first, before testing proved that the Florence sample was a piece of a meteorite, we had wondered if it was possible that the aluminum-copper alloys and the blue-green clay had formed together from natural bedrock, more specifically serpentinite. But as our investigation progressed and we learned that the Florence sample was an extraterrestrial, some began wondering if the blue-green clay had played a role in protecting the aluminum in the meteorite from oxidizing. Either way, our working hypothesis was that the blue-green clay was directly linked to the Florence sample. So we decided to focus our search efforts on areas where it could be found.

The clay from Will's Hole, which Glenn and Luca had studied, consisted of very fine grains arranged in alternating layers of blue and green. It was a match to the clay in which Valery found the Florence sample. So based on our hypothesis, we had hoped to find many tiny grains of metoeoritic minerals spread throughout the clay. But to our surprise, once Glenn and Luca began examining the material, they did not find metallic or meteoritic silicate material of any kind in either of the two layers.

I recognized that it was a significant scientific result. It not only called into question one of our basic scientific assumptions, it was also going to have immediate tactical consequences.

Perhaps it was a mistake to restrict our search to sites along the stream that contain blue-green clay.

We discussed the problem with the rest of the group during dinner that night. Chris Andronicos, with expert knowledge of structural geology, weighed in with valuable insights. After mapping the area for several days, Chris had come to believe that the blue-green clay consisted of sediment that had originally been deposited much farther up the mountain. The clay was carried downstream by a glacier that had melted in the region about seven thousand years ago, which explained how it came to be distributed all along the Listvenitovyi Stream.

I was impressed that Chris and Mike had gleaned so much so quickly after only a few mapping forays on the nearby mountains, like the one pictured below.

Chris pointed out that he was still in the earliest stage of his investigation and there were many different possibilities still left to consider. But assuming the Florence sample was once part of a meteor, as Luca, Glenn, and I believed, he could imagine at least two possibilities to explain how it might have come to be embedded in the mysterious blue-green clay.

In the first scenario, the meteor entered the Earth's atmosphere between 6,700 and 8,000 years ago. The blue-green clay was either still upstream, or recently deposited downstream by glacier meltwater. If that were the case, the clay that had traveled downstream would have still been exposed to the air when the meteor arrived. If the meteor had burst in midair upon entering the Earth's atmosphere, as many meteors do, its fragments would have been immediately lodged in the exposed blue-green clay and would still be embedded there today.

In the second scenario, the meteor could have landed upstream nearly intact less than 6,700 years ago. If that were the case, it would have slowly eroded and broken into bits over thousands of years due to weathering. Some of those bits may have been caught up in what little blue-green clay remained upstream and, together, the meteorite bits and clay could have eventually been carried downstream. Most of the bits, though, would have landed in or traveled downstream with other kinds of clay. In that case, fragments could be lodged in any type of clay deposited in the last 6,700 years.

Given the two possible scenarios, we should continue targeting blue-green clay, Chris advised. But we should also broaden our sights to include other types of clay that had been deposited at the Listvenitovyi more recently.

But without blue-green clay to guide us, how do we decide what other sites to explore? I asked myself. We might be on the verge of looking for a needle in a haystack, just as critics of the expedition had predicted.

I decided our best option was to add a new twist to the search procedure. Instead of choosing a dig site based solely on the presence of blue-green clay, we would cast a wider net and perform a series of preliminary tests before digging in, so to speak. We would obtain samples from a number of sites and examine the panned material for promising grains.

The photo above shows Sasha, Will, and Glenn getting ready to work at one of the new sites that was later labeled the Lake Hole. Based on what we found, we would then decide if a site, even one without the presence of blue-green clay, merited further attention.

That meant we would have to set up a makeshift field laboratory to sift through the hundreds of thousands of grains on a daily basis. It was not something we had anticipated, so we were not well-equipped. Our only option was to try to improvise by using Valery's portable and somewhat primitive optical microscope.

Glenn and Luca were the obvious choices to lead the laboratory effort. But after having worked with both of them over the past few years, I knew that teaming them up together might be asking for trouble. Glenn was more domineering and tended to distrust everyone else's judgment. Luca had a naturally ebullient personality, but was noticeably intimidated by Glenn's international stature and demanding standards. Unfortunately, I had no other choice but to hope opposites would attract.

Valery's microscope was crude in comparison to the state-of-the-art equipment Glenn and Luca were accustomed to working with and it was certainly not good enough to identify mineral compositions with complete certainty. But we were hoping it would be good enough to identify highly atypical specimens that would stand out from the more commonplace grains one would expect to find in a stream bed.

Glenn and Luca's mission would be to identify grains that could be a match for samples that Valery had discovered during his original expedition in 1979. That meant they would be looking for two sets of grains that seemed to have nothing in common. The first group of candidates would have a shiny, metallic appearance, like the St. Petersburg holotype. The second group would be a darker, duller, meteorite-like material, similar to the Florence sample in which we had discovered natural quasicrystals.

Glenn and Luca would also be looking for grains that did not fit either of those two descriptions but whose mineral content was indicative of the local geology. That information would be passed along to Chris, and would be useful for his study of the geological history of the region.

Thanks in large part to the detailed lab reports Glenn and Luca provided each evening, the field operations began to work much more efficiently. The team worked full blast over the next five days extracting, processing, and panning as much clay as possible.

Glenn and Luca had an amazing laboratory routine that I sometimes stopped to observe. Each day there would be five to ten plastic sacks of panned material for them to examine. The sacks were labeled with a bag number, the site location, and the date when the material had been extracted. Luca would select a bag and record all of its information in his notes. He would then carefully remove a few spoonfuls of material from the sack and empty it into a small round dish. A spoonful would typically contain hundreds of grains or more. While viewing the grains under the microscope, Luca would use a tweezer to separate each and every individual particle.

Whenever Luca found an "interesting" grain that appeared to be meteoritic or had any kind of unusual composition, he would move it to one side. Glenn would then come to the microscope and check the grains that Luca had identified and reach his own conclusions. If both of them agreed that a grain was "interesting," Glenn would put a camera over the microscope lens and take a photo. Occasionally, Valery would visit the makeshift laboratory and add his opinion. Luca would assign the "interesting" grain a number, and place it in a special vial.

The sample the team discovered the first day at the Green Clay Wall qualified as "interesting." Luca had singled it out right away. Now, it was labeled Grain #5, because it was the fifth "interesting" grain processed in the makeshift laboratory.

Once a dish of material was fully examined and the remaining grains put aside, another spoonful of material was moved from a bag to the dish for examination. It was a painstaking task. Glenn and Luca examined five to ten sacks each day. Each sack contained tens of thousands of grains.

Once a bag had been sorted through, the majority of the grains, the "uninteresting" grains, would be carefully poured back into the bag. The bag was then sealed so its contents could be taken home for more extensive study. After one bag had been completely combed through, the same meticulous procedure would be repeated for the next bag. And the next. And the next. And so on.

Glenn and Luca's work exceeded my expectations. Despite my initial concerns, their personalities blended beautifully. It seemed that working in a close environment on a shared goal brought out the best in both of them.

On one occasion, I was visiting the makeshift lab while Luca was reviewing some of the "interesting" grains from the Lake Hole, a site upstream from the Primary. I was watching as Luca stared into the microscope and a big smile suddenly flashed across his face.

"You have to see this!" he said excitedly. "We have a dodecahedron!"

A regular dodecahedron has twelve identical sides, each the shape of a perfect pentagon. Over the past few decades, synthetic quasicrystals had been known to occasionally form isolated grains with facets arranged in a dodecahedron. So finding a natural quasicrystal with twelve external facets, matching something that had already been spontaneously created in the lab, would be an important breakthrough.

Glenn rushed over to confirm that Luca was right and told us he could clearly see a half-coppery, half-silvery grain in the shape of a dodecahedron under the microscope. The external shape does not necessarily translate to the same symmetry as the internal arrangement of atoms, of course, and vice versa. But having seen a dodecahedron with a shiny metallic appearance, Glenn was also now thinking that we had just identified a multifaceted natural quasicrystal.

But when I took my turn at the microscope I had to laugh out loud. I could immediately see that we were looking at one of nature's practical jokes.

I recognized that the sample was a member of the pyrite family of minerals. Pyrite minerals include fool's gold, which novices often mistake for real gold because it has a similar color and shape. Although its atoms are arranged in a crystalline pattern with the symmetry of a cube, one of the curious properties of members of the pyrite family is that they sometimes grow facets arranged in the shape of a *distorted* dodecahedron. I like to call them "fool's quasicrystals," because each of the twelve facets has the shape of a pentagon, fooling people into believing they had found a quasicrystal. A closer examination reveals that the pentagons are not perfect. Different sides have different lengths. A diffraction pattern reveals that the atomic structure is cubic. But without knowing that, anyone could be fooled. I was quick to notice the difference because I had been collecting an assortment of fool's quasicrystals ever since I first began looking for the real thing in the 1980s.

The three of us shared a good laugh about the irony of finding a fake quasicrystal along the Listvenitovyi Stream. I just hoped we could find a real one. Even so, I thought, if a perfect dodecahedron could be synthesized in the lab it was not so outrageous to think we might *also* find it in the natural world.

AUGUST 3, 2011: *Tap-tap. Tap-tap.* A soft, insistent pitter-patter of rain on the tent was my wake-up call the next morning. The temperature had once again dropped precipitously, and I hurriedly reached for my warmest coat as I rolled out of bed.

I walked past Bucks, Olya's Russian Blue cat. Protected by a thick, double coat of fur, he seemed to take no notice of the weather. Like most mornings, he roamed around the camp like he owned the place. Will always described Bucks as acting more like a dog than a cat. He was also apparently nimble enough to be bear-proof.

Olya had another one of her substantial breakfasts waiting for us that morning with fresh caviar, jam, and hot Russian blini. The hearty food she served at each meal not only helped sustain our work efforts, it also helped fortify us against the ever-worsening cold.

The mapping team of Chris and Mike, once again undeterred by a looming subarctic storm, wanted to make one final hike to explore a distant mountain overlooking the jumble of rocks that surrounded the Listvenitovyi.

The rest of us were planning to spend our last day in the field working at various locations. Luca and Valery would be focusing on dig sites they thought were the most promising. Marina, Vadim, and Sasha would do the same at distant sites downstream, including one we called the Downstream Green Clay Wall. Unlike the rest of us, though, they were not just looking for meteorite samples. They were also looking for signs of valuable ores.

Will and I decided to gather a final round of samples from each dig site, along with samples from a few other spots we had never been able to explore.

In an effort to literally leave no stone unturned, I asked Will to climb the fifty-foot spire of rock near the Primary to retrieve a sample of clay. There was admittedly no logical explanation for that request. Just a strange idea stuck in my head from a slapstick comedy I remembered watching as a youngster called *It's a Mad, Mad, Mad, Mad World*.

In the movie, a bunch of crazy characters compete against each other to find a buried treasure supposedly hidden under a big letter "W." At one point, they run around in circles in a greedy panic, each one hoping to find the "W" before anyone else.

None of the characters ever stepped back to consider the big picture. But the movie audience could plainly see that they were all repeatedly crisscrossing the same small hill in a central area, let's call it the spire near the Primary, with four palm trees bent at different angles. It was easy for the audience to realize what the characters could not, which was that the trees formed the shape of a giant "W." The treasure they sought was hiding in plain sight the entire time.

Our expedition had always used the elevated spire as a reference point. But until that final afternoon, we had never taken the time to scramble across the stream to get a sample from the spire itself. *Who knows?* I thought.

As the morning wore on and the rain picked up, the temperatures began dropping into the low 40s and then continued to sink into the 30s, just above freezing. Everyone had returned to camp by then except Chris and Mike. They had left for their final hike right after breakfast and were now still unaccounted for. Before they left, I had insisted that they take the assault rifle along for protection.

Were they prepared for the low temperatures? Were there any bears in their path? I became increasingly worried as the hours continued to pass without any sign of them.

Chris and Mike finally marched back into camp sometime around mid-afternoon. They were both soaking wet but otherwise unscathed and well satisfied with their efforts. With their return, I felt an enormous burden suddenly lifted off my shoulders. The final day of fieldwork was now complete and, most importantly, everyone had made it through in one piece. In fact, everyone seemed very happy.

For the first time in two weeks, I could finally allow myself to relax. I am not sure anyone, not even Will, could have ever fully appreciated the sense of fear I had been grappling with. Before leaving on the trip I had heard alarming stories from colleagues about terrible incidents in the field, including fatal accidents, all of which I had kept to myself.

If anything had gone wrong during our expedition and anyone had been seriously injured, it would have been *my* fault for having organized the trip in the first place. I was the one who was responsible for putting this highly trained team of dedicated experts in harm's way, even though, truth be told, I knew in advance that there was an infinitesimally small chance of success. I do not know how I would have coped with the guilt if someone had suffered a life-threatening injury. Thankfully, I could finally put all of those concerns behind me.

That evening, our Russian hosts prepared an especially memorable feast for our final night in the field. We enjoyed our dinner outside, eating together around a large campfire. It was only August third, but the brief summer season had already taken a hard turn toward fall. We were all wearing several layers of clothing underneath our warmest coats.

Everyone was in a jubilant mood that evening, and each of us offered a toast to commemorate the adventure. Vodka had been in plentiful supply throughout the trip, and our Russian colleagues decided to hand out awards for "the non-Russians who drank the most like Russians." Glenn and Chris took top honors and were both awarded a piece of Soviet-era memorabilia—drinking flasks emblazoned with a hammer and sickle. It was strange to see that the former socialist symbol was now being used as a capitalistic souvenir. How times had changed!

Viktor ended the evening in spectacular fashion by setting off emergency flares in a dazzling display of fireworks. He handed me one of the brightly burning flares and I held the torch high for a group photo to celebrate our victory in the field. (That image is reproduced in the color insert, no. 24.) After a spate of individual photos, everyone gathered back at the campfire for what promised to be a long, boisterous evening of vodka and song.

I returned to the tent to record my thoughts in my logbook:

It has been a very successful trip by any standard. Everyone has been made comfortable and happy, even the most irascible. As a result, everyone has worked incredibly hard. I am so impressed by all of them. Unlike most geology trips that Lincoln had described for me, where some work hard, some not, and one person ends up being the camp pariah, here everyone has worked very hard. By Olya, Viktor, and Bogdan making extraordinary efforts to ensure the camp food was great and camp life was as comfortable as possible, no one, except perhaps me, has been highly stressed. Everyone has had space to express

himself or herself, even Bucks. Even if we find nothing, we will all
know that we gave our absolute maximum effort.

AUGUST 4, 2011: After a late night of revelry, some may have felt
the worse for wear the next morning. But everyone was up by 6 a.m.
to get a jump on the return trip to Anadyr. We gathered all of our be-
longings and loaded up the two behemoths for what we knew would
be a long, slow ride back to civilization. We would be trying to outrun
the rapidly worsening weather at the breakneck speed of nine miles
an hour.

Within our first half hour on the road we began spotting Kamchatka brown bears, which normally avoid human contact (see color
image 11). We kept a close watch on three of the gigantic animals
and looked out for others. We had been warned that if we saw a
small group of bears it probably meant there were many more lurking
nearby.

The occasional bear sightings continued throughout our first few
hours on the road, and at one point an especially curious bear came
within several hundred feet of our lumbering vehicles before moving
on. We were safely inside the vehicles and never in danger. But the
bear was close enough for me to fully appreciate its potential power
and to be grateful that we had never tangled with one of them.

Until now, we thought we had foolishly overpacked by bringing gloves and multiple layers of clothing. But with the temperature
continuing to drop as we left the Koryak Mountains, all of our cold
weather gear was now being put to good use. No one complained
about the increasing chill, especially not me, because it brought a
sudden end to the aggravating mosquitoes that had been making my
life miserable for the last twelve days. Finally, I could remove the hat
draped with protective netting. Finally, I could stop the application

and reapplication of DEET. Finally, *finally* this part of the expedition was over. I was celebrating all day inside my head.

As we continued our trek, I watched as a new weather system began to descend on the mountains. A chain of clouds floated from peak to peak dusting each mountaintop with a white layer of snow. I have always been an avid cloud watcher and the unpredictable cloud formations in the Koryaks had been unexpectedly fascinating. If I had ever seen the Koryak clouds depicted in a painting before making this trip, I would have just assumed they were figments of an artistic imagination. In person, the clouds and the gorgeous rainbows that often accompanied them were an inspiring natural wonder. It was like watching an ongoing, ever-changing performance, with the clouds forming and re-forming magnificent shapes the likes of which I have never seen. Wistfully, I knew I would miss being in the audience for their daily dance across the sky.

By the second day of driving, we had left the Koryak mountain range and reentered the tundra. I was once again absorbed by the natural beauty and noted with some sadness that the laughing flowers that had seemed to be mocking us when we arrived were no longer laughing. Many of the fragile white tufts were blown away by the stiff, wintry wind and only their bare flower stems were left standing. I watched the vast field of stems shiver and shake in the wind and imagined they were waving goodbye to us with every fiber of their being.

The already slow drive turned into a slog later that afternoon when the sky suddenly opened up and unleashed a hard rain. At one point, the road became so muddy that the orange behemoth got stuck in a deep depression and Viktor had to circle back to give Bogdan a tow. During the last two days of driving, the two of them had been taking turns helping each other out of the mud. Now that the rain was growing increasingly intense I began to wonder if we should continue as planned. Driving through muddy ruts in the tundra with poor visibility could be dangerous and we might easily get stuck all night in a wet, soaking mud hole.

The situation was becoming desperate when we spotted a natural gas station in the distance. It was the same station we had unsuccessfully tried to stop at during the first leg of our journey. Viktor and Bogdan slowly inched the behemoths toward the station through the mud and rain, stopping for another emergency tread repair along the way. As we got closer, Will saw the number "zero" on one of the buildings so he dubbed the site "Station Zero," although it is officially known as the Western Lakes Gas Field.

I feared the worst as we approached the station. We had not been able to call in advance about our arrival, as they had asked us to do, because our satellite phone had been knocked out of commission by the rainstorm. Previously, they had turned us away. Now, we were in the middle of a deluge and urgently needed some help.

Once we finally plodded our way up to the main building, I gritted my teeth, wondering what their reaction would be to our second unexpected arrival. Olya decided that we should try to negotiate. She took Vadim and me into the station, probably thinking that we both looked so bedraggled that the manager would have to take sympathy on us. She led us to the front desk and asked if the station could provide food and shelter for the night. At first, the manager said he was agreeable but then warned us he did not have ultimate authority to decide if the station could help us or not. The supervisor of the kitchen and rooms was the real decision maker, he told us, with what appeared to be a certain amount of trepidation.

When the much-feared supervisor finally appeared, she turned out to be a short, sweet, round-faced woman who was thrilled to welcome us, almost as thrilled as we were to be welcomed. She introduced herself as Lenechke and immediately called on her assistants to show us where we would be spending the night. I had expected she might find some empty space in the complex where we could sleep on the floor. Instead, we were shown to a set of comfortable rooms for two with heat, individual showers, hot and cold running water, and most important of all, mosquito-free indoor toilets.

The team could not believe their good fortune as they piled into the rooms. By the time we had all finished showering, Lenechke's kitchen had prepared a wonderful hot meal for us to enjoy. We were still 120 kilometers from Anadyr. But as of that evening, everyone felt like we had returned to civilization.

When we awoke the next morning, Chukotka's changeable weather had changed yet again. The heavy rain had stopped. But when we stepped outside we could see that the entire range of Koryak mountains was covered with snow from top to bottom. Winter had arrived with full force on August fifth, which meant we had barely made it out in time.

I reminded everyone about a discussion several months earlier when Valery had made his original recommendations for our trip to

Chukotka. He had told us that there was no point in going any sooner than the third week of July because the ground and river would be too cold and too hard for digging. Sure enough, even though we followed his advice and made our trip in late July, we had still struggled with the freezing soil and water at the Listvenitovyi.

Valery had also warned us that we needed to leave the Koryaks after the first week of August, or it could become too cold. Once again, his advice was spot-on. And once again, I was grateful for our extraordinary Russian colleagues.

I was hoping we could finish the rest of our trip back to Anadyr at top speed, but the overnight rain made that impossible. The tundra had become a muddy swamp and it was going to be difficult, if not impossible, for Viktor and Bogdan to make rapid progress. I remembered how intimidated I had felt when I initially climbed aboard one of their immense vehicles. My first impression was that the behemoths were invincible. Now I knew how truly vulnerable they were in the middle of the hostile terrain. It would be another twelve grueling hours of slow, careful driving before Anadyr finally came into view.

Once we saw the town in the distance, a glorious rainbow appeared over the mountains which gave the last few miles of our difficult drive a bit of a magical ending. I took a deep breath as I gazed at the beautiful setting from my usual spot in the front seat of the blue behemoth. I felt a deep sense of relief as we drove into town and safely ended the field portion of our trip.

ANADYR, AUGUST 7, 2011: After breakfast the next morning, we got right back to work and gathered for an intense scientific meeting to review everything we had learned. Chris Andronicos began by presenting many of the details he and Mike had discovered about the different types of rocks and formations in the valley and mountains surrounding the Listvenitovyi. I was enormously impressed by how much they had been able to accomplish.

Chris finished the presentation by stating his principal conclusions.

First, he could confirm with certainty that the blue-green clay in which the Florence sample was found was till from the glacier that had occupied the region eight thousand years ago near the end of the last ice age. Furthermore, there was no sign of unusual geological activity consistent with material being drawn up in a superplume or vent from deep below the surface. In conclusion, based on the observations in the field, all of the possible alternatives to the meteorite theory were now dead.

Chris had been skeptical of the meteorite theory from the start and was inherently cautious before reaching any firm conclusion. So the fact that he could now find no plausible alternative to our hypothesis was deeply meaningful to me. As I looked around the room, I could see everyone was nodding in agreement. It was a meteorite.

I had not known Chris personally before inviting him to join the expedition, and had mainly relied upon the recommendation of his former Princeton advisor, Lincoln Hollister. Having now spent two weeks with him in the field, I had come to regard Chris as a superb scientist who was remarkably talented in imagining plausible geological scenarios for a multitude of phenomena. The decision to add him to our team had paid off exponentially.

Once Chris had finished his presentation, it was up to Glenn and Luca to report the results of our digging, panning, and laboratory efforts. We had downloaded all the images they had taken of the most "interesting" grains onto an iPad, which was passed around the room so that everyone had a chance to closely scrutinize the images. The grains, numbered from #1 to #120, ranged from less than a millimeter to a few millimeters in size.

Glenn spent the next two hours reviewing the grains one by one, discussing the potential significance of each. In contrast to the optimistic enthusiasm that had followed Chris's presentation, spending two hours with Glenn going through the images of 120 samples in excruciating detail made everyone more realistic.

In conclusion, Glenn reported that in his opinion, none of the grains identified in the field appeared to resemble the original Florence sample.

Everyone in the room fell silent. I knew that Glenn tended to be pessimistic, or at least conservative, when making pronouncements. His bedside manner, so to speak, was terribly blunt. No one on the team could have been surprised at his report, because none of us ever believed that we were likely recover more meteorite grains. Even so, it was difficult to hear the bad news stated so directly. The mood in the room plummeted.

I called for everyone to place a bet: *What were the odds? What was the chance that, in all of the material we were bringing back home with us from the Listvenitovyi, there was even a single grain of natural quasicrystal?*

I walked the group through my own analysis. Figuring that we had already identified 120 of the most promising grains of interest, allowing for the fact that none of them seemed to be what we were looking for, and considering that there was a total of sixty-two bags of panned grains altogether, I estimated that there was a 0.01 percent chance of success. Less than one chance in ten thousand. Others quickly chimed in with numbers that were even more pessimistic.

Except for Luca. When he left Florence to join our expedition, Luca said, he estimated our chance of success was 0.1 percent, or one in a thousand. But now, he had decided to up the ante. He was willing to wager that we had as much as a one percent chance of success, or one in a hundred. Luca's confidence had grown tenfold for one very specific reason. He was pinning his hopes on Grain #5, the sample he singled out on our very first day in the field.

I appreciated Luca's optimism, but we both knew it was impossible to positively identify a sample based on what could be seen with either the naked eye or the images obtained through Valery's low-power microscope. We also both knew that two of our experts, Chris and Glenn, had expressed doubts about Grain #5. They were pretty sure it was not even a fragment of a meteorite, much less a grain that contained a natural quasicrystal.

After hearing all the bets, I realized that even if we adopted Luca's rosy interpretation of a one percent chance of success, everyone had just agreed that there was at least a ninety-nine percent chance that we had come up empty. That was a sobering thought.

The next morning was spent packing in preparation for the flight home. Our greatest concern was whether we would be able to get all the samples out of Russia. Chris and other American geologists had told me horror stories about samples being confiscated at the airport by aggressive Russian customs officials. Even if we managed to clear that hurdle, U.S. customs would be yet another challenge. It was illegal to bring soil into the United States. Technically, our material was "separates" and not soil. The bags of grains were perfectly legal to import because they had been panned and boiled. But we could not count on U.S. customs agents to recognize the distinction. They might decide to seize our material, anyway.

We devised a plan to give us the maximum chance of getting our samples through customs. Our team members would be taking five different routes home. So we split the sixty-two bags from our excavation sites into five sets. There would be one set for each customs route. And we would make sure there was at least one bag from each of the twelve dig sites in each set. That way, even if four of the five sets were lost to customs agents, the sole surviving fifth set would still be a representative collection of the material. Will and I would be traveling together, so we shared one of the sets. Chris, Glenn, Mike, and Luca would carry the other four. All of us planned to put the sample bags in our checked luggage.

The next day we said farewell to our Russian support staff with mixed emotions—Olya, Viktor, and Bogdan. Even Bucks, Olya's inscrutable cat, turned up to say goodbye. He was an unusual animal in that he was fully domesticated and yet thoroughly comfortable in the wild. Will had become very fond of Bucks during the trip and especially intrigued by all of his day-to-day wanderings. I watched them bond for one last time until it was time to go.

Olya gave us each a key chain in the form of Pelikan, a big-eared, fat-bellied creature that was considered good luck among the native Chukchi. She said she hoped it would bring us good luck in discovering a natural quasicrystal once we got back home. I continue to use Pelikan as my key chain today, and it is a wonderful reminder of some of the most good-hearted people I have ever met.

After a final round of goodbyes, we headed off for the airport to face our first test. The customs officials asked that we gather our bags in one place for inspection and then disappeared with all our baggage, passports, and paperwork. It was two hours before the agents finally returned and announced that we could board the Yakutia Airlines plane for Moscow. We had no idea what they had done to our luggage or the sample bags that were stuffed inside. We had to trust that they had put them on board the plane.

Once we arrived in Moscow, we waited expectantly at the baggage return and cheered each time one of our pieces came down the chute. By the end, we were relieved to see that all of them had arrived. Nothing lost thus far.

The team had to split up quickly for various connecting flights. Luca boarded a flight to Italy, Glenn headed back to D.C., and Mike was off to North Carolina. Sasha's wife and children met him at the exit to airport security. They had been waiting for him in Moscow and visiting with family while he was on the expedition. After a round of affectionate hugs, our Russian colleagues Marina, Valery, and Vadim also headed off for home.

Neither Will nor I had visited Moscow before, so we had made arrangements in advance to stay a few extra days to tour the city. Once we arrived back at the Moscow Airport to fly back to the U.S., no one ever asked us anything about the plastic sample bags spread throughout our cases. So we saw no need to volunteer any information. When our flight landed in America, we also sailed through U.S. customs without anyone asking us about the material.

The rest of the team reported that they, too, made it through customs without being stopped. None of us was ever challenged and none of our material was confiscated. Our elaborate plan for distributing the samples turned out to be unnecessary, but I had no regrets about having played it safe.

All of the samples bags would now be mailed off to Florence for Luca's review. He had been through most of the material already with the field microscope. Now, he would start the search for a natural quasicrystal all over again, combing through the millions of individual grains, but this time with the aid of an electron microscope.

Even taking the most optimistic point of view, the team had already acknowledged that the odds of finding a natural quasicrystal were extremely slim. The odds that we would ultimately fail had been pegged at ninety-nine percent.

A more realistic estimate, in my view, was much closer to one hundred percent. I had no regrets about having taken the incredible journey. But I was not about to fool myself. In my opinion, the odds of success lay somewhere on the spectrum between infinitesimally small and zero.

TWENTY

BEATING THE ODDS

I vividly recall one morning in our makeshift field laboratory when Luca jokingly asked Glenn and me what type of reward he deserved if he managed to identify a natural quasicrystal. Glenn, the wine connoisseur of the group, instantly responded:

"An expensive bottle of Château Margaux would be appropriate. And Paul," he said with a sly glance my way, "should be the one to pay for it!"

The three of us erupted in laughter at the suggestion. The truth was that finding a natural quasicrystal would be worth much more to me than a *case* of Château Margaux. But we all sensed there was little chance I would have to pay off the bet. Even Luca, the most optimistic person on our team, put our odds of success at next to nil.

The real test would have to wait until the expedition was over and Luca was back home in Florence where he could examine our samples with proper equipment.

FLORENCE, AUGUST 20, 2011: Luca began by closely examining the vial containing our 120 "grains of interest," the same grains that Glenn had already decided had nothing in common with the Florence sample. Despite Glenn's pronouncement, Luca had treated the samples like precious diamonds, carrying them home from the expedition in a special vial tucked carefully in his shirt pocket.

Unfortunately, when Luca arrived back home in Florence, he discovered a serious problem. The 120 "interesting" grains had been packed for travel during our last day in camp, the same day we had been hit hard by an unexpected rainstorm. The storm-driven winds kept knocking over the tent where Luca was working. And as a result, some of the grains were lost or damaged before they could be safely sealed in the vial.

When Luca told me the bad news, my heart stopped for a moment. *These were our most promising samples!* But Luca quickly added that the missing and damaged grains were not, in his opinion, the most important ones. For example, his favorite, Grain #5, was unaffected.

That gave me some relief, but I was now becoming concerned about the volume of work that lay ahead. The rest of the team had arrived home and would soon be shipping Luca the dozens of sample bags we had split among ourselves for safe passage through customs. I was worried that Luca was about to be swamped with more material than one person or one lab could ever hope to manage. Luca warned me that the millions of grains might take months to study depending on how much time he could reserve at his electron microscope.

Take all the time you need, was my advice. If we were lucky enough to stumble across a meteorite grain, we would need to document it carefully and treat it with the utmost care.

Luca did not have any trouble deciding where to begin his study: his beloved Grain #5. Once he started examining it under his laboratory's high-quality optical microscope, Luca realized that the photo that he and Glenn had taken in the field had not done it justice. Its most interesting side, in which many tiny metallic grains were embedded in blackish surround material, had been facing away from the camera. In fact, the more Luca studied Grain #5, the more excited he became about its potential.

Luca came up with a new way to determine the nature of the metal without damaging the sample. He would mount the grain on a stub and lean it at an angle inside his scanning electron microscope so that the electron beam would primarily hit the metal grains and not the silicate minerals surrounding them.

The good news was that Luca was working with the best high-tech equipment. The bad news was that the best high-tech equipment is always high-maintenance. The scanning electron microscope at Luca's university broke down before he could complete his study and would not be repaired for several weeks. I knew that my Italian colleague had little patience, so I somewhat expected he would find a way to jump-start the analysis.

FLORENCE, AUGUST 25, 2011: I did not have to wait very long for Luca to find a solution. Just two weeks after leaving Anadyr, a fateful email arrived from Luca with the subject line:

Chateau Margaux . . . ?? I would say yes.

I could tell without reading any further that he was referring to Grain #5. The email explained:

As it happens, turning and turning the grain at the microscope, one of the small metallic grains detached from the sample (do not worry, there are a lot of them still attached). It is a small metallic grain approximately 60 microns along. It is pure metallic, without anything attached. I washed it in acetone and then I attached it on a glass rod with glue to make a diffraction study (the only study I can do at the moment; as you know our SEM equipment is temporarily out of order). And now the news . . .

Luca had buried the lead, which was uncharacteristic of him. But he proceeded to make up for that by electronically SHOUTING the last part of the message:

. . . I saw the FIVE-FOLD SYMMETRY.

I quickly clicked on the X-ray diffraction image attached to the email. I leaned forward in my chair as the image appeared and my eyes popped. It was a deceptively simple image that was full of meaning. *Could this be real?* I thought. It seemed too good to be true.

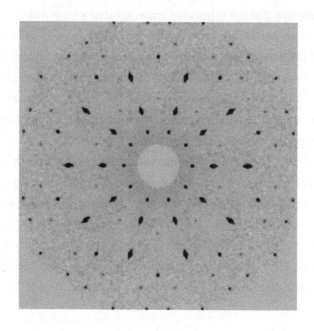

The image showed unmistakable evidence that the atomic arrangement in Grain #5 had the impossible five-fold symmetry that can only be found in a quasicrystal. And in this case, unlike the Florence sample, there was no mystery to solve to determine who had found the sample, or where or when. There could never be any doubt about its origin. Because we had personally witnessed its discovery.

I could have shouted for joy. I could have burst out of my office at Princeton and told everyone I ran into what I had just seen. I could have fired off emails to Lincoln Hollister and the expedition team. I could have phoned Will to share the amazing news. But I did not want to do any of those things. Eventually, but not yet. I wanted to stop and fully absorb the historic moment. This was not supposed to have happened. None of us thought it would. But there was the image, right before my eyes. It was a profoundly emotional experience and personal milestone.

I sat and stared at the image, thinking about each of the talented members of the expedition team, most of whom had been total strangers to me before our journey. They had happily volunteered their time, energy, and well-being to join our quixotic quest. Christopher Andronicos. Vadim Distler. Marina Yudovskaya. Sasha Kostin. Michael Eddy. Valery Kryachko, the man who set all of the events in motion decades ago when he found the first sample of khatyrkite.

I thought about Luca, with whom I had been communicating almost every day for the last four years. Glenn MacPherson, whom I had first met two and a half years earlier at the doorstep to the Smithsonian Natural History Museum. And my son Will, whom I first encountered when he was stark naked and a whole lot shorter.

I thought about the fact that Grain #5 had been discovered at the Green Clay Wall thanks to Marina's suggestion; dug out by Will and Sasha; carefully panned by Valery; identified on-site by Luca;

immediately confirmed on-site by Valery, Luca, and Will; reexamined in the makeshift lab by Chris and Glenn; and then reviewed by the entire expedition team. The success this day was a testament to the diverse group of people who had committed themselves to our mission.

It was amazing to me that any of them agreed to make the long trek to the Listvenitovyi, considering how inhospitable and remote it was, especially because we all suspected the trip would most likely end in failure. More amazing still was that everyone gave their absolute utmost every second of the journey, never stopping to complain or question their sacrifice. Luca and I would now have the pleasure of informing all of them that their dedication had just been repaid fivefold.

Lincoln Hollister owned a big piece of this accomplishment, as well. He had not been able to join the expedition, but had been a vital participant in the search for natural quasicrystals ever since our first meeting in January 2009, shortly after Nan Yao and I had spotted the first electron diffraction pattern with five-fold symmetry in one of the tiny grains of the Florence sample. Lincoln was my mentor as I planned and prepared for the expedition, sharing insights and advice drawn from his distinguished career. I could already envisage that great big smile of his when he heard the news.

I thought about our donor Dave, who had funded our expedition. He had been especially generous after I warned him that the expedition might yield nothing and that the cost had increased by more than a factor of two. Dave doubled down and supported us without hesitation. I began to imagine how I would let Dave know that his gift was about to return great, unexpected scientific dividends.

There were many other people who had contributed mightily to the decades-long story, beginning with Dov Levine, who had been there from inception. It felt like yesterday that we were working together, but it had actually been nearly three decades since the two of us had first developed our theory of quasicrystals. We had proven that it was theoretically possible for the atomic structure of matter to have five-fold symmetry, as was later confirmed in the laboratory. That had led to the development of three-dimensional models of the new form of matter with Dov and with Joshua Socolar. Now, almost thirty years later, we were in a position to prove that nature had beaten all of us to the punch, making the first natural quasicrystals billions of years ago.

I thought about Caltech's Ed Stolper, who, at a critical moment, gave me the impetus I needed to go forward and who also pointed me toward two more heroes, John Eiler and Yunbin Guan. Princeton's Ken Deffeyes challenged me to follow my instincts to search for *natural* quasicrystals and introduced me to his protégé, Peter Lu. All of them would be thrilled and would appreciate the groundbreaking consequence of this discovery, as would Roger Penrose and David Nelson, who helped inspire the original ideas that started me down this path.

None of this would have happened if Richard Alben had not first introduced me to the study of atomic structure, or if Praveen Chaudhari had not encouraged that interest. And then, of course, there was Richard Feynman, who made me fall in love with physics in the first place.

There were, in fact, hundreds of scientists around the world, too many to name, who had applied their theoretical and experimental talents to help establish a new branch of physics.

How could I possibly thank everyone? I wondered.

All of these thoughts were flying through my mind, almost simultaneously. I finally forced myself to push myself away from my desk and walk down the hall. I thought a cup of coffee might help me clear my head before attempting to compose the first round of announcements.

When I returned to sit down, though, I was once again frozen in place by the amazing X-ray diffraction image on my computer screen. I took a long, slow sip of coffee and felt myself slipping into another grateful reverie. Here was a substance older than the Earth itself but fresh with possibilities.

TWENTY-ONE

L'UOMO DEI MIRACOLI

WASHINGTON, D.C., OCTOBER 5, 2011: "I recognize this," Glenn said proudly as the image popped up on screen, *"Allende!"*

I laughed, knowing that Glenn was teasing me and making fun of himself at the same time. History was repeating itself but with a happy twist. A few years earlier, Glenn had used the same microscope to examine a similar image but had erupted with fury and contempt.

The image that had angered him was from the powdery material Luca discovered at the bottom of a vial labeled "Khatyrkite," which was recovered from a former colleague's secret laboratory. Glenn was convinced material from the famous Allende meteorite had been put in the vial, blaming "a capricious, if not overtly malicious, God" for the mix-up.

A lot had happened in the two and a half years since that episode. Luca had once again sent Glenn a sample to review, but this time there was no question in Glenn's mind that it was legitimate. After all, he had been part of the expedition that had recovered it from the original dig site.

So Glenn was delighted, rather than enraged, to find the close resemblance to Allende in the new sample. He had fully embraced the notion that our natural quasicrystal was, incredibly enough, a piece of an Allende-like meteorite. What he was showing me on the screen was striking new proof.

We were looking at a cross section of Grain #121, the second promising candidate Luca had identified since returning home. I had made the three-hour drive from Princeton to Washington to be with Glenn when he took the first high-resolution images.

To the untrained eye, Grain #121 looked like a big clod of mud surrounded by a finely grained gravel. But the innocuous-looking clod and all of the material around it were packed with important information about the birth of the solar system.

"That is a chondrule," Glenn said, "the oldest part of a chondritic meteorite, dating back more than 4.5 billion years." That assessment alone told us the sample was legitimate. "The added material surrounding the chondrule is called the matrix," Glenn explained, "which is normally composed of certain characteristic minerals."

Chondrules and matrix are the two major components of a carbonaceous chondrite meteorite. The same microscopic materials are found in Allende. So it was significant that we were now observing the same features as Allende in the new grain.

To prove the point, Glenn began gleefully exploring Grain #121. Using his electron microprobe, he measured the chemical composition at various spots in both the chondrule and the matrix. Chondrites contain a complex mix of minerals. Before taking each measurement, Glenn would predict what the composition would be, based on his extensive research experience with the Allende meteorite. He was right every single time.

"Anyone would see this image as classic Allende," he concluded.

Just then, a more junior scientist in his group happened to pause at his office door and glance at the image on the screen. Glenn asked if she could identify the image. "Allende, of course," she responded, as if it were a stupid question. Glenn and I smiled at one another with satisfaction as she continued along her way.

Upon closer inspection, Glenn found something important in Grain #121 that he had never found in Allende: tiny white chips. With respect to the chondrule, two of the chips are located at roughly four o'clock and six o'clock, as seen in the image below.

Since the chondrule itself had been cut in half during the sample preparation, the image suggested that the chips of material had been embedded in the chondrule, which was made of silicate, when it formed more than 4.5 billion years ago. That indicated that the white chips were also likely to be more than 4.5 billion years old.

When Glenn checked the chemical composition of the curious

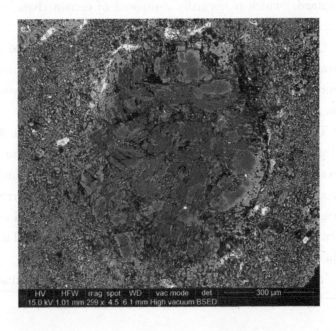

white chips with his electron microprobe, he became even more animated.

"The chips are cupalite!" Glenn announced excitedly. We looked at one another and half laughed and half cheered. "Unbelievable!" I shouted.

Cupalite, along with khatyrkite, was one of the metallic alloys found in the original museum samples that had made everyone, including Glenn, dismiss them as fake. Finding cupalite chips buried inside the chondrule, the oldest part of the meteorite, was the most direct evidence yet that the meteorite's aluminum-copper alloys formed in space 4.5 billion years ago when our solar system was in its infancy.

FLORENCE, OCTOBER–NOVEMBER 2011: When I circulated the news about Grain #121 to the team, effusive praise and congratulations for Luca began to pour in from everyone, including our anonymous benefactor, Dave. They were all beginning to recognize that Luca was, indeed, *L'Uomo dei Miracoli*, as I affectionately called him. *The Miracle Man*.

New discoveries kept coming. *L'Uomo dei Miracoli* struck again with Grain #122. I had been sending the team regular bulletins about Luca's progress. Now, I was issuing another update for the third time in less than six weeks:

> It seems hard to believe that only 10 days have passed since Report #2 and there is already important news to report: Icosahedrite has been discovered in a third grain recovered from the Koryaks, this time from a different location: Will's Hole.

My report emphasized that the three meteoritic grains that were now making scientific history had been found in three different locations along the Listvenitovyi. Marina and Sasha had recovered Grain #5 from the Green Clay Wall; Grain #121, the grain with the chondrules that had intrigued Glenn, came from the Downstream Green Clay Wall; and the latest discovery, Grain #122, was recovered from Will's Hole. Only the last site, Will's Hole, had blue-green clay.

The fact that the samples came from different types of sites distributed over hundreds of meters was informative. It meant that all of the sample bags from all of our dig sites could be potential sources of meteoritic material, not just the bags that came from sites containing blue-green clay. That meant Chris Andronicos had been right when he advised us to broaden our search to other areas. I was once again reminded that Chris was a crucial member of our team. Without him, we might never have retrieved so many promising samples.

I thought Luca's discoveries would be especially gratifying news for everyone who had continued to work so diligently during the expedition, battling the nearly frozen clay and often plunging their bare hands into freezing water after all of our shovels had broken.

Then ten days later, I sent out yet another report describing yet another miracle from *L'Uomo dei Miracoli*. Grain #123, also from Will's Hole, was significant because it contained the largest grain of icosahedrite seen to date that was in direct contact with meteoritic material.

The importance of such a finding cannot be overstated. Before the expedition, Lincoln and Glenn had bemoaned the fact that we could not find indisputable evidence that the quasicrystals in the Florence sample were in contact with meteoritic minerals. We needed that evidence, they said, in order to help prove our case that the quasicrystals had formed naturally. Now, with Grain #123, we finally had a large and unambiguous piece of evidence.

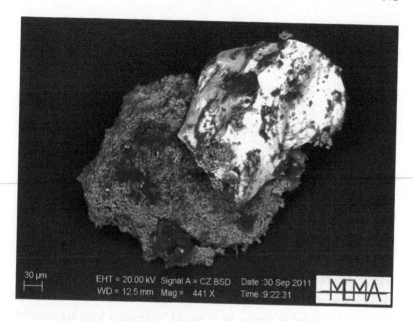

In the image above, the quasicrystal, icosahedrite, is the chunk of material in the upper right. It is physically connected and enmeshed with the silicates below.

The miracles continued as Luca combed through the bags. He soon identified three more grains—#124 from Will's Hole and #125 and #126 from the Primary trench—all of which appeared to be meteoritic. With his incredible string of discoveries, Luca's title was now firmly established. *L'Uomo dei Miracoli!*

PASADENA, FEBRUARY 2012: The grains Luca had found were sent off to Caltech for John Eiler and Yunbin Guan to analyze with the NanoSIMS.

By now, based on our earlier tests of the Florence sample, our leading hypothesis was that the aluminum-copper alloys had formed in space and arrived on Earth as part of a carbonaceous chondrite meteorite. But we were always on the lookout for contradictory information. If we were to ever find an example in which icosahedrite was attached to ordinary terrestrial minerals, that single exception would be enough to force us to rethink our entire theory. Everything we thought we understood about the extraterrestrial origin of icosahedrite would be thrown into doubt. So it was crucial to repeat the oxygen-isotope NanoSIMS test on the silicates found in each and every sample recovered from the Listvenitovyi stream.

Luca was working in overdrive, so I had become accustomed to receiving fast, accurate results. Shifting focus to Caltech's spectrometer, though, was like being forced into the slow lane. The equipment was always booked several months in advance with a backlog of important research waiting to be done. It was also constantly breaking down and in need of repair. So it would be another long six months before we would begin to get the first reports.

Once the results finally began trickling in, they proved without a doubt that the silicates had exactly the same oxygen isotope ratios as the Florence sample, which, in turn, had the same signature as a classic CV3 carbonaceous chondrite.

John Eiler was one of the geologists who had initially warned me not to make the expedition. There was essentially a zero chance, he said, of finding any more meteorite samples. But the best scientists always enjoy being surprised and discovering something unexpected. So despite his initial doubts, or perhaps because of them, John was excited to send me the good news that proved he had been mistaken.

HOUSTON, TEXAS, MARCH 2012: One month after the Caltech measurements were completed, Glenn shared our phenomenal results with other meteorite experts throughout the world at the annual Lunar and Planetary Science Conference (LPSC). The team was confident Glenn would be our best representative. He was well known and well respected in the LPSC community.

Glenn went to the meeting armed with all of the impressive evidence we had collected both before and after the expedition. He met with the Nomenclature Committee and made the forceful case that we had discovered a new meteor impact. It was crucial for us to obtain their official imprimatur in order to convince the rest of the meteorite community that our finding was legitimate. But gaining such acceptance is often an uphill battle. The committee is notoriously ultraconservative and super-finicky.

Glenn must have bowled them over because the committee immediately accepted his case that the grains were meteoritic. They also agreed to assign the meteor the official name we proposed: Khatyrka, in honor of the river we had managed to drive and float across in our two behemoths.

Five months later, Glenn took charge of drafting the first scientific paper about the expedition results, which was published in the prestigious journal *Meteoritics & Planetary Science* (*MAPS*) on August 2, 2013. The paper combined valuable contributions from each member of the expedition, as well as from Lincoln Hollister, John Eiler, and Yunbin Guan. Some members of the team had moved to other institutions since the expedition. Our team was now spread around the world in Florence, Boston, Moscow, Washington, D.C., Houston, West Lafayette, Pasadena, Johannesburg, and Princeton.

We were anticipating a fair amount of skepticism because of the unusual aluminum-copper alloys found in our samples. So Glenn had made sure that our *MAPS* paper was absolutely meticulous. It was full of images and exceedingly detailed, including a description of the pristine clay layers in which the samples were found and the exhaustive, quantitative measurements of mineral compositions and isotope concentrations that had been taken from each sample.

The paper, entitled "Khatyrka, a new CV3 find from the Koryak Mountains, Eastern Russia," established the existence of a new meteorite and provided new evidence for the natural provenance of a number of aluminum-copper metallic minerals, including the first-known natural quasicrystal, icosahedrite.

The meteorite community had no trouble accepting the conclusion that the silicates in our samples were meteoritic. The oxygen-isotope tests were unambiguous proof of that, as far as anyone was concerned. But as always, it was the natural origin of icosahedrite and the other aluminum-copper minerals found in the Khatyrka meteorite that were more difficult for some to accept. Like other types of geologists, meteorite experts had always been taught that minerals with the symmetry of an icosahedron were impossible. So were the strange metallic aluminum alloys we were reporting. Nothing like this had ever been seen before in meteorites. Glenn's presentation at the LPSC meeting and the *MAPS* paper marked the beginning of a discussion about quasicrystals and metallic alloys that would last for years.

Despite the assemblage of evidence in our scientific paper, a few meteorite scientists remained vocal skeptics of our conclusions. We never criticized any of them for doubting our claims. After all, Lincoln and Glenn had initially reacted the same way when Luca and I brought them our initial discovery in 2009.

Most of the doubts tended to melt away whenever we were given the opportunity to present our remarkably thorough test results. But for those who never seemed to take the time to learn about all of the detailed evidence, our conclusions were considered impossible. They preferred to stick to their view that quasicrystals and alloys containing metallic aluminum could not be made by any natural process whatsoever, either on Earth or in outer space.

Three years after our paper was published the discussion was still raging in certain pockets of the scientific community. So Glenn, being a colorful debater, decided to engage in more public discussion. He prepared a poster presentation for the 2015 Lunar and Planetary Science Conference, which was attended by ten thousand scientists. Glenn stood alongside the poster at the convention and personally explained important details about all of the evidence we had collected since his first presentation in 2012.

To make the evidence even more accessible, the team prepared a handout to accompany Glenn's poster with a list of FAQs and answers. In a typically theatrical flourish, Glenn challenged anyone to come up with a plausible alternative explanation for all of the evidence we had collected to prove the existence of a natural quasicrystal and metallic alloys.

The only group that attempted to respond to the challenge was a team of Russian geologists who had attempted their own expedition to Chukotka after hearing about our success. They prepped for the expedition by meeting with our Russian colleague, Valery Kryachko. Even so, their trip to the Listvenitovyi Stream turned out to be an utter failure. They never found a single grain of meteorite, quasicrystal, or aluminum-copper alloy in any of their panned material.

Instead of questioning their own methodology, the team responded to Glenn's challenge by publishing a paper asserting that, despite mountains of documentation, our findings were mistaken. The metal alloys in our samples had to be synthetic, rather than natural, they claimed. They floated the idea that our samples had been accidentally created by gold miners as a result of the dynamite blasts they set off while loosening clay to pan for gold. The blasts, the Russian team proposed, could have shredded nearby tools or piping or some other unknown mining equipment composed of aluminum alloys. Then, bits of that metallic material could have been propelled into nearby rocks at high speed. Included in the rocky targets, they supposed, were the remains of a CV3 carbonaceous chondrite meteorite, which, just like the well-known Allende meteorite, did not originally contain any metallic alloys. Their conclusion was that the accidental fusion of exploded man-made metal with an ancient meteorite created our samples.

While imaginative, the concept does not hold up under scrutiny.

First, the Russian team could not produce a single metal implement used by gold miners with the right chemical composition to explain either the quasicrystals or the aluminum-copper alloys we discovered in our sample. In fact, while investigating the use of aluminum-copper alloys, I had discovered that metals with such compositions are far too brittle for any kind of practical use. It is true that aluminum-copper alloys with only a few percent of copper added to aluminum, or vice versa, are commonly used. But the alloys found in the Khatyrka meteorite, which included 50-50 or 60-40 mixtures of these metals, have no known industrial applications for one simple reason: They are too brittle.

Second, if the blast proposal were true, one would expect to have found metal alloys fused with ordinary terrestrial minerals. Terrestrial minerals are exponentially more abundant than meteoritic material at the Listvenitovyi. In fact, even before the Russians proposed their blast idea, we had been searching systematically for such examples to test our own meteorite hypothesis. We never found an example, and neither did the Russian team. Not a single one.

Third, the Russian blast idea did not account for the quasicrystal grain totally encased in stishovite that Luca had found in the Florence sample. Stishovite is a silicate that can only be created at ultra-high pressures. Those pressures could never be created by metal shrapnel propelled by a dynamite blast.

Since it could not have been created by the blast, the stishovite must have been part of the meteorite before the blast, according to the Russian team's logic. The metallic aluminum alloy, which the Russians claimed was synthetic, would have then been propelled by a blast into the meteorite. But if that had been the case, the stishovite grain, which would already be part of the meteorite, according to the Russian hypothesis, would have a large hole where the synthetic alloy had pierced through, and there was absolutely no evidence of that.

Fourth, the blast hypothesis could not explain why some of our grains were found in pristine clay buried deep under the surface, which had apparently sat undisturbed for thousands of years. No dynamite blast could have sent bits of metal from nearby tools hundreds of meters downstream and through so many layers of sediment in a claybed, especially not without leaving a lot of other obvious damage in the area.

In the end, these and other weaknesses we pointed out in the Russian team's explanation made it clear how strong our case for a natural origin was and how difficult it was to find any plausible alternative.

Our team would have much preferred that the Russian scientists had succeeded in finding additional meteoritic samples, since that would have provided more scientific data. But I had always known it would be difficult for any other group to duplicate the success of our expedition because they could not hope to replicate the single most important component of our success: the people on our team.

Others could dig and pan as much as we did, but they would never have diggers as committed or as careful as Will, or a panner as experienced and as skillful as Valery. They could never have a meteorite expert as qualified as Glenn. They could never hope to duplicate Valery's, Marina's, and Vadim's decades of experience working in Kamchatka and other areas with natural ores. They would never need their own mapping team to study the geological history of the area, because Chris and Mike, with support from Marina and Sasha, had already done all of that hard work for them. And perhaps most importantly, they would never find anyone with nearly the knowledge, talent, and all-out, no-holds-barred dedication as Luca.

I am especially proud of the fact that our team has kept its scientific standards sky-high and has constantly challenged its own conclusions to avoid becoming overconfident or careless. Lincoln Hollister has been a model for us all in this regard. More than any outside individual or group, we have always been the toughest critics of our own work. We question and challenge one another over and over to make sure no details or theoretical possibilities are missed.

In the years since the expedition, we have methodically eliminated all of the various possible explanations for how our samples could have been created by natural terrestrial forces or accidental industrial or mining activities. But there has always been a nightmare possibility that we have returned to over and over again: *Could we be the victims of an elaborate ruse?*

The NanoSIMS oxygen-isotope measurements confirmed that the silicates were from a CV3 carbonaceous chondrite meteorite, dating back to the birth of the solar system. But the NanoSIMS could not be used to test the metallic alloys because the alloys contained no oxygen.

Could a devious person combine genuine Allende-like meteoritic material with synthetic aluminum-copper alloys, expose the mixture to some combination of high pressure and temperature and produce samples like those we recovered?

The first problem we ran into while exploring that outlandish scenario was the same one that made the Russian blast idea so untenable. There are no readily available metals with the same compositions of aluminum and copper that were found in our Khatyrka samples. The alloys are simply too brittle to ever be of any industrial or commercial use. Fakers would have had to synthesize the peculiar metal combinations on their own, beginning with pure aluminum and copper. They would have had to engineer that process before 1979, when Valery recovered the first samples from the Listveni-tovyi. A problem with that particular timing, of course, was that it would have been several years before Dov Levine and I had considered the possibility of quasicrystals and before they had been discovered in the laboratory. So that would mean there would have been no motivation for creating metallic alloys with such peculiar chemical compositions. But assuming the faker did so anyway and mixed them with real meteoritic minerals, he would have had to place the fruits of his devious labor in an obscure stream in the remote Koryak Mountains and bury them deep in thick clay, not knowing if anyone would ever discover them.

While all of that was ludicrously improbable, we nevertheless went through a brainstorming exercise to see if we could design a procedure to create the type of grains we had observed without producing any tell-tale signs of fakery. Try as we might, none of us was ever able to come up with anything close to a workable theory.

We did come up with our own version of a fanciful idea, though, which sounds like something out of a *Star Trek* movie.

Imagine that the Khatyrka meteor was the result of a collision between an ordinary carbonaceous chondritic meteor and an alien spaceship. One could then imagine that the never-seen-before combination of aluminum-copper metal in the Khatyrka meteor might be a remnant of that spaceship. It has always been a fun explanation for us to fantasize and joke about, especially because it would mean that our quasicrystal was ultimately proof of life on other planets.

Of course, all this is in jest. The point of the laughable alien spaceship theory is that, as crazy as it sounds, it is harder to disprove that theory than any of the more plausible possibilities we considered, all of which we have successfully managed to test and disprove.

But if the alien theory was a joke, what was the real secret of how and when our natural quasicrystal formed?

TWENTY-TWO

NATURE'S SECRET

In less than a year after returning home from Russia, our team had already obtained an overwhelming amount of new evidence. We had proven beyond a reasonable doubt that nature made quasicrystals long before humans fabricated them in the laboratory, and that the examples we had recovered were not of this world. They were visitors from outer space.

We could have stopped there, declared victory, and moved on to other research. But neither my nor Luca's DNA permitted that. Our curiosity was more stoked than ever and we were totally committed to finding out *where* our meteorite originated, *when* it formed, and *how* it was created. There was no simple way to answer all those questions. The only way forward was to try everything. Simultaneously.

Leave no stone unturned. That had been my mantra ever since the first natural quasicrystal was discovered in a long-forgotten museum sample. In the wake of our expedition that all-out approach was more apropos than ever.

Dig out every detail from the natural samples we had brought back. Design experiments to reproduce the extreme conditions in outer space so we could test our theories using man-made alloys. Identify new ways to find the original source of the Khatyrka meteorite.

Collect and investigate meteorites similar to Khatyrka for natural quasicrystals or for other supposedly "forbidden" metallic aluminum alloys. And finally, figure out how to do all of those things at the same time because no one could ever know in advance how long any of them might take or which, if any, of these ideas would turn out to be the most fruitful.

As a result, our research efforts since 2012 have been unusually diverse, with novel and occasionally risky experiments. New teams of scientists, each with an array of highly specialized knowledge, have been enlisted to help pursue our quest. We have had our share of painful failures along the way. But what remains most impressive to me is the extraordinary progress and remarkable insights we have realized in such a short amount of time.

ALUMINUM WORMS AND MINERAL LADDERS

We began with Grain #125. Of all the grains we recovered from the Listvenitovyi, it had the longest and clearest example of a contact between an oxygen-bearing silicate and khatyrkite, the crystal aluminum-copper alloy that is the most abundant metal in our samples. Studying the textures near the contact seemed like a promising approach to take in order to try to understand the powerful forces that created the unusual combination of minerals.

One of our earliest team members, Lincoln Hollister, was the ideal person to lead the investigation. Lincoln and I began working together in January of 2009, just a few days after our initial discovery of a natural quasicrystal. He was renowned for his ability to piece together the history of rocks based on their structure and composition, which was exactly the kind of analysis we needed. Lincoln had officially retired from Princeton the same month we began our expedition but insisted he had no intention of withdrawing from the project. He loved the challenge of being at the forefront of trailblazing research.

Our first new team member would be Chaney Lin, a graduate student who had come to Princeton in the fall of 2011 to study theoretical physics with me. Once he was exposed to the mysteries and puzzles surrounding natural quasicrystals, he was hooked. Just like the rest of us.

Chaney started off with a summer project aimed at finding new meteorite samples in the dozens of bags of material we brought back from Chukotka. Luca had already completed two full passes through the hundreds of thousands of grains, so it seemed a good time to enlist a fresh set of eyes. Chaney's long-term goal was to become a theoretical physicist, which involves more math than microscopy. So before he could examine any grains to see if they had the right chemical composition, he needed to learn how to use an electron microscope, which is a delicate business in its own right.

With tutelage from Nan Yao, director of the Princeton Imaging Center, Chaney soon became one of the best electron microscopists on campus. He had both the patience and skill to extract precise and meaningful information from our tiny materials. By the end of his summer assignment, Chaney and another graduate student had completed a third pass through all the material. They discovered two additional meteorite samples, which was a great cause for celebration.

Chaney then decided to keep working on our investigation, in addition to his work on theoretical physics. As an undergraduate, he had spent the last four years on the East Coast at New York University. But having grown up in Los Angeles, Chaney retained an ample amount of the laid-back demeanor one often finds in a Californian. Among his many positive traits, he could accept criticism without becoming defensive or emotional. He would always listen to my feedback with a receptive smile before responding with thoughtful and creative comments. I decided he would be an ideal protégé for Lincoln, who had earned a reputation on campus as a wonderful, but demanding, mentor.

When I introduced Chaney and Lincoln, seen below, the two of them hit it off instantly and *voilà!* They dove into analyzing every minute component of Grain #125, beginning with contacts between the silicate and the khatyrkite.

Chaney soon made his first major scientific breakthrough. Using an electron microprobe to study Grain #125, he determined that the wormy threads in the khatyrkite metal were nearly pure aluminum, which was something never seen definitively in any mineral before. Finding that impossible substance, along with the impossible metallic aluminum alloys, greatly enhanced the mystery of the Khatyrka meteorite. Chaney presented the evidence of pure aluminum to Lincoln and me in his usual, understated manner. But it was evident from his ear-to-ear smile that he was bursting with pride at the discovery.

Lincoln expertly interpreted Chaney's image, shown below. He pointed out that the regular texture of the dark, wormy aluminum threads between the channels of whitish khatyrkite—a mixture of one part copper and two parts aluminum—was a sure sign that the metal grain had somehow completely melted and then rapidly cooled.

If the original liquid mix was one part copper and slightly more than two parts pure aluminum, Lincoln said, it would naturally separate as it cooled and solidify into thick strips of khatyrkite with thin wormy threads of excess aluminum, which was exactly what we were seeing in Grain #125.

Studying the silicate material, the darker substance on the other side of the metal-silicate contact, with the electron microscope was more problematic. When Chaney and Lincoln first viewed it in the scanning electron microscope and checked its chemistry with the electron microprobe, they found an unusual composition and texture that they could not readily identify. With Lincoln looking on, Chaney

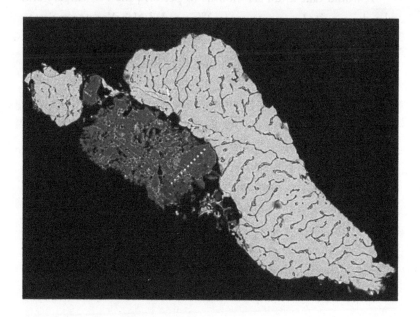

worked for many weeks and used a barrage of creative techniques to solve the mystery. Nothing worked.

The two of them finally decided the problem was that the composition varied wildly across the microscopic space of just a few interatomic distances. The microprobe could only report the average composition over a much greater area, which had the effect of blurring fine-scale variations. We needed to find another experimental approach that could resolve differences in composition occurring over very short distances.

After consultation with Luca Bindi and Nan Yao, we devised a plan using a special piece of equipment known as an FIB—a focused ion beam. It would be a risky surgical operation that we had not used on any of the other grains so far. The FIB would cut and remove an ultrathin slice from the puzzling area of the sample. Then the slice would be studied with a transmission electron microscope, which, unlike the microprobe, was powerful enough to measure compositional differences over very short distances.

It would take a full six months to perform the FIB surgery and take the required measurements. We needed Nan Yao's expertise to prepare the sample. First, he carefully reviewed it with Chaney, Lincoln, and me. Then, he painstakingly deposited an extremely narrow strip of platinum on top of the tiny sample in a predetermined place that appeared to have the greatest compositional variation. The location is seen in the previous image along the dotted line. The total width of the platinum strip Nan applied was less than one-hundredth of the thickness of a human hair.

The sample was then sent to FIB expert Jamil Clarke at Hitachi High Technologies in South Carolina. He would focus an intense ion beam on the sample and blast away material surrounding Nan's tiny platinum strip. The platinum was thick enough to repel the ions, so the slice of material that lay directly beneath the metal was sure to remain intact.

The ion beam created a depression all around the platinum strip.

Within the depression was a gossamer-thin wall of meteorite material, which was left standing in the microscopic crater like a fragile butterfly wing. With exquisite care, Jamil detached the delicate piece from the rest of the sample and shipped everything back to us.

The nearly transparent slice was barely visible when we opened the package. One sneeze, I thought, and the sample would be lost. Once we were able to examine it under the transmission electron microscope, we understood why the microprobe had never been able to get a clear read on its composition and texture. Instead of being a uniform layer composed of a single mineral, it looked like a complex, microfine mess. And *that* revelation opened the door to another series of important discoveries.

The slice was originally composed of silicate material that would normally be contained in the matrix material commonly seen outside the chondrule of a carbonaceous chondrite meteorite. But there was one significant difference. In this case, the image showed that the silicate material had melted and then rapidly cooled. It felt like the story was beginning to come together, because that was consistent with the wormy aluminum threads we had already found in the other portion of the grain, which also indicated that it had been melted and then rapidly cooled.

Because the cooling of the silicate had happened so quickly, the microfine mess that was revealed by the transmission electron microscope caught the violent, ancient process in suspended animation. The liquid had formed rivers and streams between remnants that had not melted, and each of the streams had rapidly solidified into a texture that looked like a ladder (as shown on the next page).

The white rungs on the ladder consisted of amorphous silicon dioxide, a glassy substance. Even more significantly, the dark rungs on the ladder consisted of a rare mineral called ahrensite. Just like stishovite, which had been found in one of our other samples, ahrensite can only form at ultra-high pressures. Chaney and Lincoln determined

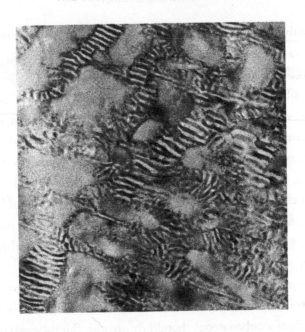

that the pressure must have been at least 50,000 times the normal atmospheric pressure on Earth. Temperatures would have had to have reached at least 2,000 degrees Fahrenheit in order to melt both the aluminum and copper.

As we continued to study the rest of the silicate in Grain #125 beyond the FIB slice, we recognized that it was composed of minerals arranged in shapes reminiscent of the loose matrix in the grain Glenn and I had reviewed together shortly after returning home from the expedition. The difference was that, this time, the matrix mineral grains were crushed together in a dense lump, which was exactly what one would expect if it had undergone a high-velocity impact with another asteroid in space. The impact would generate a shock wave that would squeeze and compress the loose matrix material into the shapes we were seeing under the microscope. And it would have melted the matrix in certain places where the temperature and pressure were especially high. With the discovery of ladders composed of ahrensite and silica and the observation of crushed matrix material, we now had explicit, quantitative proof that the Khatyrka meteorite had undergone one of the most powerful impacts ever detected in a CV3 carbonaceous chondrite.

Everything we had learned so far was further confirmation that the Khatyrka meteorite was exceptional. Luca and I were more energized than ever and ready to pursue the next open questions.

Were the natural quasicrystals already part of the Khatyrka meteorite when it began forming in the nascent solar nebula 4.5 billion years ago? Or were they created at a later time as a result of a collision?

Lincoln favored the second theory, which was that the natural quasicrystals formed after an intense impact. He found it more likely that aluminum and copper were chemically bonded to more typical chondritic meteorite minerals beforehand. Some of the minerals would have melted due to high pressures and temperatures caused by the impact, he theorized, freeing their atoms to form both the quasicrystal and the two "impossible" crystalline aluminum-copper alloys, khatyrkite and cupalite, which were found in the sample.

Glenn MacPherson's favorite theory, on the other hand, was that the quasicrystals and aluminum-copper alloys had existed from the start. He believed it was more likely that the pure aluminum and copper had condensed directly from the solar nebular gas during the earliest stages of the solar system and had been part of Khatyrka all along.

It was not immediately obvious how we were going to distinguish between the two theories. Luca and I needed to come up with a different kind of experiment. *But what?* we wondered.

LOST IN SPACE

My philosophy of leaving no stone unturned occasionally caused problems.

While Chaney and Lincoln continued to search for more clues in Grain #125, Luca and I looked for a new way to study our samples. We desperately wanted to find new, nondestructive tests. Samples from the expedition were a very limited resource, and we wanted to preserve as much material as possible for multiple rounds of testing.

The process used to prepare our samples for the electron microscope was becoming a problem. We had to first embed the sample in a special holder filled with hot epoxy; then let it cool; and finally cut through the encased material to expose a smooth surface that could be studied.

The solidified epoxy helped keep the sample intact during slicing but introduced a problem unique to our material. The heat of the epoxy tended to fracture contacts between the metal and silicate. Our samples were especially vulnerable because of the difference in thermal expansion rates between aluminum-copper alloys and silicates. We were trying to study contacts between those materials and needed them left as undisturbed as possible.

A promising alternative was X-ray tomography, essentially a CAT scan for minerals. The test can identify minerals within a sample and produce an extraordinarily useful three-dimensional reconstruction. It had already become a well-established technology for human medical diagnostics but was still a relatively new technique for studying minerals. It could not approach the fine resolution we had already achieved with the FIB experiment and was also not as precise as the electron microprobes we were already using. But it offered one major advantage: It would not require the destructive hot epoxy and slicing procedures.

Luca and I had both read about the emerging technology and decided to arrange a trial experiment. Luca was able to obtain access to a low-resolution machine. So he tested part of a sample that had not been epoxied. The results looked promising, so I arranged to perform more precise scans at the University of Texas High-Resolution

X-ray Computed Tomography Facility, which had some of the best machines in the world. All I needed to do was provide the lab with clean samples that had not been epoxied.

The only grains that remained unaltered at this point were two of the samples Luca was working with in Florence. So those were the grains that would be sent to Texas. Luca carefully packed up the two samples, Grain #124 and Grain #126, using the same methods he had been using to send me samples for the previous five years. He hand-carried the padded box to the Air Express office in Florence, as usual, and mailed it off to me in Princeton.

And then, nothing. That was the last we ever saw of it. Air Express lost all trace of our priceless, tiny grains.

I was aghast. Absolutely, positively horrified. Our expedition team had beaten the odds and traveled thousands of miles to the easternmost edge of Russia, traversed the tundra, crossed the turbulent Khatyrka River, evaded enormous Kamchatka brown bears, battled relentless mosquitoes, dug up tons of nearly frozen clay in freezing water with our bare hands, fought our way back to civilization through storms, conveyed our sifted material out of Russia, painstakingly sorted through millions of grains only to have some incompetent, faceless person misplace two of our most valuable finds?

For the next few months, I kept a semi-frantic watch on my mailbox as Luca hounded Air Express.

Did the package make it out of Italy? Was it stuck in customs, baggage claim, or buried in the back of a delivery truck? What about the computerized tracking system?

Increasingly desperate, Luca tried to enlist the shipping company's help by explaining how exceedingly rare the grains were, how difficult they had been for us to obtain, and how important they were to scientific research and our understanding of the fundamental nature of matter.

Luca became progressively agitated as time wore on. But the Italian

Air Express office just shrugged off the incident. They never managed to figure out what happened to our package. Worst of all, the staff never seemed to care.

A year earlier, I had questioned the wisdom of using express mail, after Glenn had used it to send me some rare samples from the Smithsonian. But he just laughed, and told me everyone in geology uses some form of express mail, even when dealing with the most valuable minerals. I got the same answer from John Eiler at Caltech. Everyone convinced me that I was being paranoid.

But then, the Air Express disaster happened. And suddenly, nobody was laughing anymore.

From that point on, I refused to entrust any delivery service with our Khatyrka samples. Nothing would ever be sent by express mail again, not even international packages to Luca in Italy. I insisted that everything be delivered by hand, if not by me, then by a student or a colleague who happened to be traveling to and from Italy, California, Washington, D.C., or Princeton.

Unfortunately, the lost samples were also the *last* samples we had left that had not been epoxied, so we were never able to perform an X-ray tomography test, the 3-D imaging experiment that might have opened up a whole new dimension in our study. It was, and remains, a major disappointment. But we are still considering using the technique to sort through additional meteorites in search of more metallic aluminum alloys and quasicrystals.

QUASICRYSTALS UNDER PRESSURE

We had to accept the fact that two—*TWO!*—of our most valuable samples were lost. We tried to move on as best we could, and turned our focus to finding new ways to determine how Khatyrka and its natural quasicrystals formed.

Evidence from Grain #125, along with earlier studies, indicated that the Khatyrka meteorite had experienced a high velocity collision in space, the impact of which created ultra-high pressures. That raised an important question: Could we ever expect a quasicrystal that was buried in a meteorite, specifically icosahedrite, to survive extreme pressures of more than 50,000 times the atmospheric pressure at the surface of the Earth?

If not, we would know that icosahedrite could never have been part of Khatyrka during the birth of the solar system, the theory Glenn favored, because it would not have survived the high velocity impact Khatyrka later incurred while traveling through space. Instead, we would know that it must have been created sometime after the last big impact experienced by Khatyrka when the pressure was much lower, as Lincoln believed.

That issue struck at the core of our research. The stability of quasicrystals and the interatomic forces that hold their atoms together are questions of fundamental importance for condensed matter physicists and materials scientists. Stability tests had already been performed at lower pressures or temperatures, but no one had ever performed tests at the combination of high pressures and temperatures relevant to Khatyrka. Decades earlier, however, Dov Levine, Josh Socolar, and I had constructed cardboard and plastic models suggesting that there could, in principle, be interatomic forces to ensure stability under extreme conditions.

This time, there was no need to put any of our actual samples at risk. The test could be performed with man-made icosahedrite quasicrystals. The fact that synthetic quasicrystals had become so easily obtainable was a reminder of how long I had been fascinated by the material. It was astonishing to think that quasicrystals were now so commonplace that man-made versions could be purchased inexpensively from a chemical company.

Arranging for the high-pressure, high-temperature stability experiment itself was much more challenging. Very few labs are capable of performing such delicate tests with reliable accuracy. Luca identified Vincenzo Stagno and his colleagues Ho-Kwang Mao and Yingwei Fei at the Carnegie Institution for Science in Washington, D.C.

The setup required three components: a tiny tungsten carbide "anvil" cell measuring less than an inch across for producing the pressure; a particle accelerator nearly three miles in circumference that can accelerate electrons to speeds that are 99.9999998 percent of the speed of light and bend them in circles that cause them to emit high-intensity X-rays; and advanced magnets and detectors that can aim the X-rays very precisely at material within the diamond cell and measure the X-ray diffraction pattern that is produced.

Accelerators and detectors like this only exist at five places in the world. The Carnegie Institution has a dedicated high-intensity X-ray beamline at Argonne National Laboratories outside Chicago, which was where the trial experiments were performed. The final measurements were done at a similar facility called SPring-8 (Super Photon ring-8 GeV) in Hyogo Prefecture about 250 miles southwest of Tokyo.

Our plan called for surrounding synthetic samples of icosahedrite, the same type of quasicrystal we had discovered in Khatyrka, with a graphite heating device and placing it in the tungsten carbide anvil cell, a box whose walls could be squeezed together by a press to crush whatever was inside. As the pressure and temperature gradually increased, the X-rays emitted by the electron beam were aimed at the quasicrystal so that any changes in the diffraction pattern could be continuously tracked. The exquisite measurements took one and a half years to plan and execute, and the results were well worth the effort.

The findings were conclusive and indisputable. Icosahedrite did not transform, not even under the extreme conditions of pressure and temperature that Khatyrka experienced during a high-velocity impact.

This meant that, in principle, icosahedrite could have been part of Khatyrka since its inception over 4.5 billion years ago, as Glenn had proposed, and subsequently survived all the impacts the meteorite experienced in space. Even so, the findings were not enough to prove Glenn's theory was correct. Lincoln's alternate explanation was still a possibility, as well. It was conceivable that the crystal metal alloys and the icosahedrite could have formed as a direct result of an intense impact in space. Icosahedrite might still turn out to be the direct result of an impact.

NOBLE GASES

We knew that parts of the Khatyrka meteorite dated back 4.5 billion years and that sometime after that there was an intense collision in space between Khatyrka and another meteor. But *when*?

To address that issue, we needed to perform another extraordinarily difficult experiment with yet another group of highly trained specialists. Luca brought tiny bits of silicate from the Khatyrka meteorite to Henner Busemann, Matthias Meier, and Rainer Wieler at the Swiss Federal Institute of Technology in Zurich, shown left to right in the photo on the next page. Wieler had specifically designed the laboratory to measure rare helium and neon isotopes in meteorites. Matthias and Henner, his protégés, performed most of the experiments. Matthias was particularly captivated by the project and volunteered to lead the test.

Helium and neon are known as noble gases, two of the six elements in the right-most column of the periodic table that are odorless, colorless, and have very low chemical reactivity.

As they travel through space, meteoroids are bombarded by

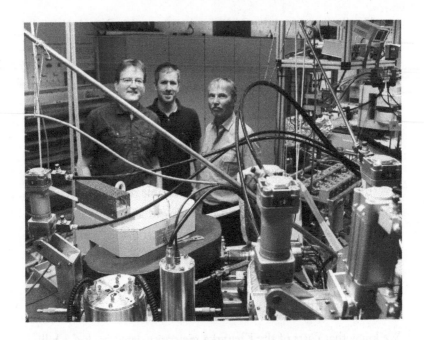

cosmic rays, energetic subatomic particles traveling at nearly the speed of light. The cosmic rays strike atomic nuclei in the rock, creating helium and neon isotopes with different numbers of neutrons than the helium and neon nuclei typically found on Earth. By measuring the percentage of atypical nuclei, they can gauge how long a meteoroid had been exposed to cosmic rays in space.

If the Khatyrka meteorite had experienced a strong impact in space, all of its accumulated helium and neon would be lost as a result of the elevated pressures and temperatures created by the collision. If it then continued its journey through space, cosmic ray bombardment would resume and create a new population of atypical helium and neon isotopes. That process would continue for as long as Khatyrka remained in space. Once Khatyrka reached its ultimate destination and landed on Earth as a meteorite, the Earth's atmosphere would protect it from any further bombardment.

Matthias would begin by destroying the sample to extract the isotopes. What made the experiment so difficult was that he would then have to trap and isolate each and every helium and neon atom that emerged. Next, he would measure the concentration of those isotopes.

When I visited the Swiss laboratory, it occurred to me that the sophisticated equipment, with its maze of crisscrossing pipes and tubes, looked a lot like a plumber's nightmare. Once the sample was vaporized, the equipment would capture the gas that was created and transport it through a series of twists and turns, which were specially designed to ensure that only the helium and neon would survive the labyrinth. The microscopic survivors would be counted and categorized by the detector at the far end of the tubing.

It took several years to set up, perform, and analyze the results of the highly delicate procedure. It was a calculated risk, because the sample would have to be destroyed in order to extract the isotopes. Fortunately, the gamble paid off magnificently. The Zurich test revealed nuanced information about Khatyrka's history that we could never have obtained otherwise, and helped us create a timetable for its journey through space.

Caltech's NanoSIMs test had already established that some of the minerals in Khatyrka dated back to the birth of the solar system, about 4.5 billion years ago.

Then, sometime between a few hundred million and a billion years ago, according to Zurich's isotope test, Khatyrka was part of a large parent asteroid that underwent a powerful collision. The impact was sufficiently violent to kick out all of the helium and neon isotopes that had been created by cosmic rays up to that time. There may have been earlier major collisions, but this was the most recent one, based on isotope measurements recovered from the tiny sample.

For the first time, we could estimate the date of the collision that had probably created the stishovite and the ladder of ahrensite and silica we had observed in our samples.

The results also showed that the Khatyrka fragments were part of a meter-sized chunk that had broken off from its parent asteroid between two and four million years ago. Some event, perhaps a gentle collision with another asteroid orbiting the sun, caused it to detach and begin its slow and meandering path toward the Earth. Based on Chris Andronicos's earlier evaluation and carbon dating, we knew that the chunk entered the Earth's atmosphere about seven thousand years ago.

The windfall of information was staggering. The results proved that the meteorite's impact on Earth could not have been responsible for the stishovite and ahrensite. That collision was simply not powerful enough. If it had been, there would have been no rare helium or neon isotopes whatsoever detected in our sample.

These results were independent confirmation of what we had claimed all along. If the impact on Earth was too gentle to get rid of the helium and neon isotopes, and therefore not strong enough to create the stishovite and ahrensite found in our sample, it could not have been powerful enough to create the aluminum alloys we observed in Grain #125. The only logical possibility that remained was that the metallic alloys were already part of Khatyrka *before* it had entered the Earth's atmosphere. They were made in outer space and melted at some point during Khatyrka's earlier travels through the solar system.

This was one of those cases where *Leave no stone unturned* really paid off. When Luca and I first considered trying these difficult noble gas isotope experiments, we were concerned about having to sacrifice bits of our rare samples on a risky test that might yield nothing. But sticking to our philosophy, we had charged ahead despite the long odds and were rewarded with more information about Khatyrka's history than we could have ever imagined.

A NAMESAKE

I was already impressed with everything we had learned about Khatyrka, but then came a new series of miracles provided by our *L'Uomo dei Miracoli*, Luca Bindi.

By now, we had given up fighting with Air Express and had accepted the permanent loss of Grain #124 and Grain #126. But Luca had been keeping a secret from me. Small chips from Grain #126, each about the thickness of a fingernail, had broken off while he was packing the samples for shipment. The main piece had gone to its demise at the hands of Air Express. But Luca had recovered the small chips and stored them in a tube in his lab.

When he finally had a chance to take a look at the leftover chips, Luca found something unusual. Most of the other grains included metallic minerals of aluminum and copper, but Grain #126 also had metallic minerals containing aluminum and nickel. Luca soon discovered a crystalline mineral containing roughly an equal mix of aluminum, nickel, and iron that had never been seen before in nature.

As with all of our other new mineral discoveries, Luca meticulously prepared a proposal for the International Mineralogical Association. This time, however, he chose to hide everything from me. Luca had privately decided to name the new mineral "steinhardtite" in my honor. He consulted with other members of the expedition team, who secretly approved and agreed to coauthor the application. Not even my son Will, who joined the conspiracy, told me what was going on. Luca submitted the paperwork to the International Mineralogical Association and soon thereafter, steinhardtite was officially approved.

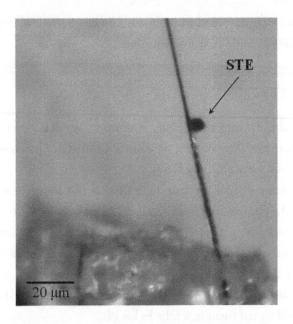

I was deeply touched when Luca told me the news. Such a thing is a rare occurrence and a true honor, particularly for a theoretical physicist. It was especially meaningful for me that the whole thing was engineered by my teammates. Thanks to them, I am forever mineralized.

The amount of natural steinhardtite available today is microscopic. The holotype sample, the tiny grain shown suspended on a thread in the image above, is now permanently housed at Luca's Natural History Museum in Florence. A similar sample sits in a treasured box on my desk at Princeton.

A SECOND QUASICRYSTAL?

And then, *L'Uomo dei Miracoli* did it again. While trying to recover more steinhardtite from the microscopic chips of Grain #126, Luca discovered something even better—a second kind of natural quasicrystal. If someone had not known the story up to this point, they would no doubt say that finding two different kinds of natural quasicrystals in a single sample was impossible. But by now, we were used to the fact that virtually everything we were achieving was impossible.

The second quasicrystal was different both chemically and geometrically from the first natural quasicrystal, icosahedrite. Chemically, the new quasicrystal was a mix of metallic aluminum, nickel, and iron, similar to steinhardtite, but with different percentages of the three elements.

What was absolutely stunning about the new quasicrystal was its symmetry. Just as there can be crystals with different symmetries, we knew that there could, at least in principle, be natural quasicrystals with different symmetries. But none of us had ever expected to see a natural quasicrystal with a different symmetry in the same meteorite.

Khatyrka was turning out to be an absolute marvel.

The first-ever natural quasicrystal discovered several years earlier, icosahedrite, has six different directions along which one can observe the famously forbidden five-fold symmetry. The second natural quasicrystal, though, had only one direction with forbidden symmetry. And it was forbidden ten-fold symmetry.

As shown on the top panel to the right, the

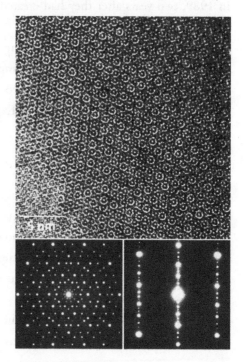

structure is full of little rings of atoms that form decagons. The diffraction pattern on the bottom left panel of the previous page confirms the ten-fold symmetry along one direction. But other directions are periodic, like an ordinary crystal, as proven by the regularly spaced rows of diffraction spots on the bottom right panel.

Finding a completely different type of quasicrystal was far, far beyond anything Luca and I could have ever imagined. Over Skype, we cheered our good fortune.

Once again, Luca submitted the evidence to the International Mineralogical Association with a proposal for a new mineral. They rapidly voted in favor and accepted our proposed name of *decagonite.*

Decagonite is a new mineral, but a familiar substance to quasicrystal experts. A quasicrystal with the same composition and symmetry had been synthesized by An-Pang Tsai and his collaborators in 1989, two years after they had created the world's first bona fide example of a synthetic quasicrystal.

No one had ever anticipated finding a decagonal quasicrystal in nature. But that was the feat Luca accomplished, all from a tiny, leftover chip from the long-lost Grain #126. Imagine what my talented colleague might have discovered if Air Express had not been so careless with the rest of the sample.

THE AMAZING GRAIN #126A

Incredibly enough, Luca managed to squeeze a third discovery out of the remnants of Grain #126. One of those chips turned out to be so important that it was given its own designation. We called it Grain #126A, and it was chock-full of new evidence about the Khatyrka meteorite.

Since the onset of our investigation, we had been looking for a sample in which metallic aluminum was in direct contact with, and chemically reacting with, the silicates normally found in carbonaceous chondritic meteorites. The best example we had managed to find up to this point was in Grain #125, which Chaney and Lincoln had been studying. Unfortunately, the grain's mineral contacts had been broken during the epoxy process.

We got an unexpected surprise with Grain #126A, shown in the image below.

At first sight, it appears to be another example of a dog's breakfast, the memorable phrase Glenn had once derisively used to describe the messy images we had recovered from the broken remains of Luca's computer hard drive.

Here, too, the image looks like a jumble. But at the microscopic level there is incredibly informative detail. The investigation of this single little scrap of a dog's breakfast managed to occupy our team— Chaney, Lincoln, Luca, and me—for more than two years. At key moments, we reached out to colleagues from our expedition, Chris

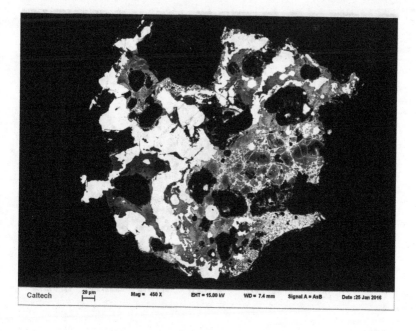

Andronicos and Glenn MacPherson, for guidance. We ultimately recruited even more specialized experts for the team from Caltech.

In the sample, one can immediately identify many examples of the metallic minerals, which are the whitish and light gray materials. The silicate and oxide minerals are represented by the dark gray materials. Most importantly, we could tell by this image that the two materials had chemically reacted with each other.

A prime example of this can be seen in the small section of the dog's breakfast magnified below, which I call the "turkey." The bird's head and beak are in the upper left quadrant and the plump round turkey body is in the middle.

The turkey represents a region where metal and silicate melted and reacted with each other due to an impact, likely the major collision the Zurich isotope test had identified as happening hundreds of millions of years ago. All along the boundary between the metal and the silicate was a thin layer full of mysterious round beads which were nearly pure

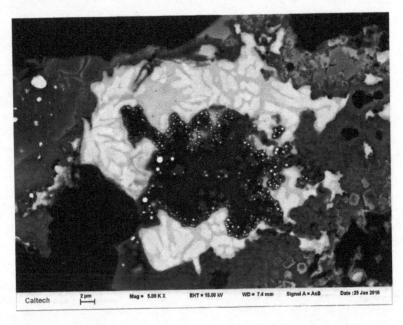

iron. There was also a delicate arrangement of nonmetallic spinel crystals, which were oxides containing aluminum and magnesium.

It was the first example of this configuration of minerals we had seen among the Khatyrka samples. The spinel and iron beads were the product of a rapid heat-producing chemical reaction that took place when the aluminum in the surrounding metal came in contact with the silicate's oxygen, magnesium, and iron. The aluminum atoms joined with the magnesium and oxygen in the silicate to form the spinel; the iron liberated from the silicate condensed to form the beads.

But what caused that chemical reaction? And how could we be certain?

CANNONS AND TURKEYS

I decided we needed to try to find the answers experimentally. Iron beads have been observed in lunar samples that experienced impacts on the moon's surface. So I wondered if impacts in outer space might account for the iron beads in Khatyrka, even though they have a completely different chemistry from lunar surface material.

Hoping to find a way to test my idea, I spent the next few months holding conversations with various experts who had written about the iron beads found in lunar samples. They, in turn, pointed me to other experts, who pointed me to other experts—the usual sort of laborious, time-intensive search for information that had characterized the entire quest.

While I was speaking with an engineering professor at Caltech, a familiar name surfaced. He told me that a colleague, a geophysicist named Paul Asimow, had once studied the formation of iron beads during high-velocity impacts. *Bingo.*

My youngest son Will had been a geophysics major at Caltech. During his undergraduate years he had introduced me to Paul, a professor he greatly admired. Paul was thin and wiry and a bundle of energy.

He was also brilliant, creative, and intensely curious. Once he had an idea for an experiment, he was lightning quick to act.

Paul, at left, had access to the Caltech research laboratory that contained a rare piece of equipment called a "propellant gun," which is basically equivalent to a specialized cannon. The cannon itself is about five meters long and works just like a traditional cannon. The front end of the 20-millimeter-bore cannon, known as the breech, is loaded with gunpowder along with a two-millimeter-thick projectile composed of a hard, rare metal called tantalum. At the other end of the cannon is a custom-designed target consisting of a stack of synthetic or natural materials embedded in a stainless steel chamber, which is about three inches wide and equally thick. The particular materials used for the stack vary depending on the experiment being performed. The chamber containing the stack of materials is attached with nylon screws at the far end of the cannon, where the target assembly is encased in a large rectangular "catch" box.

When the cannon is fired, the projectile travels at about 3 times the speed of sound and generates a shock wave that passes through the target stack and lasts for less than a millionth of a second. At its peak, the shock pressure duplicates the pressure that Khatyrka experienced in space. The force of the impact shears the nylon screws, sending the steel chamber flying into the rectangular catch box at the rear of the assembly, where it is later recovered and taken apart for study.

My first email to Paul included an image showing just a small section of our iron beads in the turkey, asking if he had ever seen anything like it. He responded right away, excited because he had previously used the gas gun to study the formation of iron beads in stacks of various *synthetic* metals. Here was a *natural* example of the same phenomenon he had studied. It was immediately fascinating to him.

We soon began to discuss how to perform a test by putting together a stack of materials that may have been part of Khatyrka prior to its high-velocity impact, as established by the Zurich isotope test. We were hoping we could re-create the formation of iron beads by smashing the stack with the tantalum projectile from his cannon.

I had thought about trying a cannon experiment several years earlier. But at that time, it had not been clear what kind of materials should be included in the stack. By the time we discovered the iron beads in

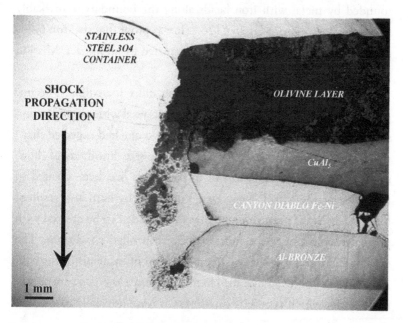

Grain #126A, we knew much more about the composition of Khatyrka. Using that information, Paul designed a test using a target consisting of various layers, as shown on the previous page. The first layer would be olivine, a typical meteoritic silicate, followed by a synthetic copper-aluminum alloy, then a natural iron-nickel alloy from the Canyon Diablo meteorite topped off with a synthetic aluminum bronze. Everything would be held together tightly in the stainless steel container.

Once the shot was fired, the impact part of the experiment was over in an instant. But after removing the sample from the catch box, we spent many additional months carefully dissecting its contents to determine what happened. We were hoping to prove that the impact could produce the iron beads that we had found in Grain #126A. But that would turn out to be the least of our discoveries.

A series of reactions had occurred at the side of the container when the impact created a shock wave that propagated through the stack. Incredibly enough, there was a tiny region of silicate surrounded by metal with iron beads along the boundary, remarkably similar to the turkey in Grain #126A. It was proof that the iron beads observed in Khatyrka could have been created by an impact. Mission accomplished.

But the test revealed something even more incredible. The impact had created grains of icosahedral quasicrystal with a composition similar, but not identical, to icosahedrite. No one had expected that.

More than thirty years after their discovery, hundreds of thousands, perhaps even millions, of quasicrystals had been created in labs all over the world. They were known to be hard and resilient but it had always been thought that they needed to be created with great care and under the most strictly controlled conditions. The violent smashing together of a combination of minerals in the cannon shot was nothing like the chemically pure, low-pressure conditions under which synthetic quasicrystals were normally made.

1 μm

It was another incredible step forward in the investigation. And our unexpected success inspired a series of shock synthesis experiments with Paul's cannon. One of them was designed to see if we could produce decagonite, the second natural quasicrystal that had been discovered in our samples. For this, we changed the mineral compositions in our stack to include nickel, an ingredient of decagonite.

This, too, was a success. The impact produced a series of flower-like arrangements, as shown in the image above. The grayish petals are decagonal quasicrystals. Most astonishing was the bright white substance that formed in the middle of the flowers. That, curiously enough, has the same composition as steinhardtite.

The shock experiments were now so successful that they began taking on a life of their own. Occasionally, they created quasicrystals and other crystals with compositions that had never been seen before, either in nature or in the lab.

That result has led Paul Asimow and me to consider using the gas gun to collide many other combinations of elements together, which will be a new and exciting way to search for new materials. We might be able to find examples of quasicrystals with particularly useful combinations of physical properties, including strength and electrical conductivity. Or we might discover materials with a different kind of orderly arrangement of atoms that has never been contemplated before.

THE MOST AMAZING QUASICRYSTAL YET

The iron beads were just one of several surprises we discovered in Grain #126A. By carefully identifying each type of mineral and taking note of which was connected to which, we were able to reconstruct in detail what had happened during the enormous impact Khatyrka experienced hundreds of millions of years ago.

In particular, we now began focusing on whether icosahedrite and the other aluminum-copper minerals were produced during the impact or whether they had existed beforehand. Despite all of our testing, neither possibility had yet been ruled out.

To get the answer to that question, we first had to determine if there was any icosahedrite to be found in Grain #126A. Chaney had spent weeks searching through the complicated islands of metal in the sample. Almost all of the metal minerals that he found were either crystalline khatyrkite or other aluminum-rich phases. He could never find any icosahedrite. But we were not about to give up.

As a last resort, we sent Chaney to Pasadena to work with mineralogist Chi Ma, who had access to an electron microscope with even finer resolution than the one Chaney was working with. Chi soon found a tiny fleck of metal that had been much too small for Chaney to resolve. And incredibly enough, it revealed a remarkable combination of metallic alloys in contact with icosahedrite.

Now we could finally report that, within the very same grain of material, one could see examples of icosahedrite alongside evidence of the chemical reactions between metal and silicate. I was thrilled because I knew that the finding truly cemented our scientific discovery. There could no longer be any question that silicate and metal existed together in space and experienced the same physical conditions, adding yet another type of direct proof that our quasicrystals were made in space.

The fresh round of discoveries also included three new crystalline minerals composed of different combinations of aluminum, copper, and iron that had never been observed in nature before. All three have now been officially accepted by the IMA. The three new minerals are named hollisterite, after my Princeton colleague Lincoln Hollister; kryachkoite, after our Russian colleague Valery Kryachko; and stolperite, after Caltech's former provost Ed Stolper, who provided me with crucial insights and encouragement in the early days of our investigation. Ed also paved the way for me to work with several of the highly accomplished scientists in Caltech's geophysics department.

The most remarkable of all the new minerals discovered to date in Grain #126A is designated with the provisional name "i-phase II" (our proposed official name is "quintesseite"). It is indicated by the arrows in the image on the following page. It forms little ellipsoidal shapes that are arranged like petals on a flower and surrounded by a complex arrangement of other minerals. Another section of Grain #126A had looked like a turkey to me. This one looked like a barking dog. Its head is top center, facing right, with its mouth open in mid-bark.

The discovery of i-phase II represents the completely unanticipated *third* natural quasicrystal to be found in the Khatyrka meteorite samples.

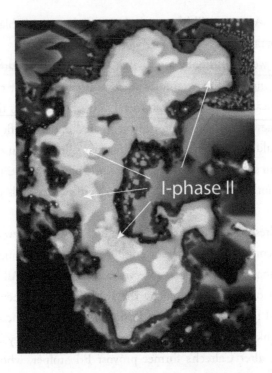

The provisional name, i-phase II, signifies that it is the second icosahedral quasicrystal phase of matter. Just like icosahedrite, the third natural quasicrystal has icosahedral symmetry and is composed of the same elements—aluminum, copper, and iron. But it has a distinctly different mix of those three elements, thus making it chemically and structurally distinct.

By analyzing the shapes of the icosahedrite and i-phase II and the minerals surrounding them, Lincoln and Chaney were able to fill in some of the remaining blanks about what had happened to Khatyrka hundreds of millions of years ago. They determined that the tiny fragment of metal containing i-phase II had liquefied as a result of the impact and then solidified to form the complex of metallic alloys seen in the barking dog image. That meant i-phase II had definitely formed *after* the impact. On the other hand, the configuration of icosahedrite and the metal that surrounded it showed that they had definitely not been melted by the impact. That meant icosahedrite definitely existed *before* the impact.

After __and__ before the impact? How could both things be possible?

The answer appears to be that the massive shock that Khatyrka experienced produced incredibly wild variations in pressure and temperature. Within a few millionths of a meter, which is approximately the diameter of a red blood cell, there are certain regions of material that melted and certain regions that did not. As a result, Khatyrka included two different quasicrystals with icosahedral symmetry that are composed of distinctly different combinations of the same elements and that formed at different times. It was a truly astonishing discovery.

An important implication was that we now knew for sure that that icosahedrite, the first natural quasicrystal we ever identified in Khatyrka, predated the impact. That was consistent with Glenn's theory that it may date back 4.5 billion years to the beginning of the solar system and disproved Lincoln's idea that it formed after impact.

From my point of view, the discovery of i-phase II has been the most significant discovery to date for another reason. It is the landmark discovery I have been hoping for ever since 1984, the year that my student Dov Levine and I first published our theoretical proof. That was when I first began flirting with the idea of searching for a natural quasicrystal by scouring display cases at prominent mineral museums.

My goal had always been two-fold. First, I wanted to prove that quasicrystals were stable enough to have formed in nature, as I long suspected. Secondly, I wanted to know if finding a natural quasicrystal could open the door to discovering types of quasicrystals that had not been known before.

With the discovery of i-phase II, my dream came true. For me, it is more important than any of the other natural quasicrystals we have discovered because it is the first one found in nature *before* being synthesized in the laboratory.

Scientists have barely scratched the surface when it comes to learning the unique properties and potential applications of quasicrystals. More than a hundred different compositions have been synthesized in the laboratory over the last three decades. But most are closely related chemically to the original quasicrystals discovered by Dan Shechtman and An-Pang Tsai.

The lack of variety is due to the fact that there is no theoretical guidance for deciding which particular combinations of atoms and molecules can make this unique and fascinating form of matter. Finding new examples is usually done by trial and error. The simplest approach adopted by many scientists is to make a small change in the chemical composition of a synthetic quasicrystal that is already known to exist.

But that limits the possibilities. If one is interested in finding quasicrystals with properties that are more interesting, both from a practical perspective and a scientific point of view, one can improve the odds by searching to see what nature has made without human intervention. Toward that end, Paul Asimow and I are currently planning more cannon experiments. Experimenting with new methods of fabrication will be another way to advance the science.

Despite all our success, there is still a huge question about Khatyrka that remains unanswered and continues to intrigue me.

By some mysterious process, nature has somehow managed to form quasicrystals with metallic aluminum in direct contact with nonmetallic minerals rich in oxygen, despite the fact that aluminum has a voracious affinity for oxygen. For reasons we cannot yet explain, the aluminum in our natural quasicrystals did not react with the nearby oxygen in the silicate. Normally, the chemical forces would be sufficient to cause the oxygen to react with the aluminum to make corundum, an extremely hard version of aluminum oxide. If we could understand nature's process, it might teach us a new, more efficient way to make both ordinary crystals and metallic aluminum-bearing quasicrystals.

PHOTONIC QUASICRYSTALS

But do we have any indication that any quasicrystal might have novel and useful properties for science and industry?

Yes, we do. We can either simulate quasicrystals on a computer or create artificial samples using a 3D printer, as shown below. The example pictured here was constructed in 2005 at Princeton by Weining Man and Paul Chaikin, who collaborated with me to study the "photonic" properties of quasicrystals.

The study of photonics is directly comparable to electronics.

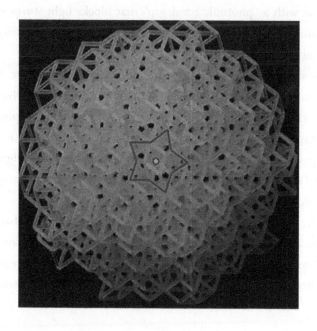

Electronics involves the passage of electrons through materials. Photonics involves the passage of *light waves* through materials. If we could replace electronic circuits with photonic ones, the speed of transmission would be increased and the heat loss due to resistance would be reduced. One of the challenges is to find a way to use photonics to reproduce the effects of semiconductors like silicon, germanium, and gallium arsenide. Those are the materials that comprise transistors and other electronic components used for the amplification and transmission of signals in computers, cell phones, radios, and televisions.

The defining property of a semiconductor is that electrons are totally blocked from propagating through it if their energy lies within a certain band of energies. Engineers take advantage of the so-called "electronic band gap" to control the flow and the information carried by electrons.

Something similar exists in photonics. It is possible to make a material with a "photonic band gap" that blocks light waves within a certain band of energies. The first examples were photonic crystals, introduced and developed twenty-five years ago.

By shining microwaves through our 3D printed structure, Weining Man, Paul Chaikin, and I have shown that quasicrystals have some of the same features as photonic crystals. They also have photonic band gaps. Most importantly, the band gap properties of quasicrystals are superior to those of photonic crystals because they have higher rotational symmetry. That makes their photonic band gaps more spherical, which is advantageous in practical applications.

The photonic quasicrystal example illustrates the point that there may be advantages to quasicrystals over regular crystals in some applications due to their distinctive symmetries, provided we can find examples with the right combination of chemistry and symmetry. We may hit upon good examples by trial and error in the laboratory, but now, we can also imagine discovering useful examples in nature.

IMPOSSIBLE?

A key to finding how natural quasicrystals formed in nature is knowing *when* and *where* they were created. So far, our intensive studies of Grains #126 and #126A have only managed to take us part of the way toward answering those questions for the Khatyrka meteorite.

Based on our cannon experiments and our studies of Grain #126A, we know that i-phase II formed as a direct result of an impact that incurred hundreds of millions of years ago. That impact shocked, heated, and melted a combination of metals that cooled and solidified to create i-phase II and the distinctive configuration of metals surrounding it.

On the other hand, we have also observed that the quasicrystal icosahedrite was not melted by the impact. So it definitely existed beforehand, perhaps long before the great collision in space. That leaves us with many questions yet unanswered. How and when did it form? Was it the first quasicrystal to form in the solar system? Is it common or rare? Is the currently favored hypothesis correct—did it truly form in the early solar nebula? Were there lightning storms in the nebular dust cloud, as some of us have speculated, that aided the formation of aluminum-copper alloys? Or is it possible that the quasicrystal is part of a "pre-solar grain," formed during the demise of an older star system and traveling through space to join our solar system? Whatever the case, what other novel minerals were made? And what effect did all of this have on the evolution of our solar system?

Although we continue to pursue many different experimental approaches, as of this writing, nature still has the answers to all of those questions under lock and key. Perhaps there is more to be found in further studies of the Khatyrka meteorite. Or perhaps someone will find examples of aluminum-copper alloys in other meteorites to provide further clues.

But if I were to dream the wildest dream about where to find the right key to open the next scientific door, it would be to visit Khatyrka's parent asteroid.

Khatyrka, like most meteors, was at one time part of a much larger parent asteroid that is still orbiting the sun. Sometime between two and four million years ago, long after the great impact, Khatyrka broke off and raced away from its parent like an errant toddler. Eventually, it got lost in the Earth's atmosphere and either exploded in midair or crashed intact on the Earth's surface.

If we could locate its parent asteroid, land on its surface, collect samples, and study the chemical and isotopic compositions of all the minerals they contained, the origin of Khatyrka would be revealed.

The sobering thought, though, is that there are approximately 150 million potential asteroid parents longer than a football field that are lying in the asteroid belt orbiting the sun. If one were to include smaller asteroids, the list would be much longer. So it would obviously be *impossible* to find Khatyrka's parent—or even a close relative—among the huge crowd.

But you might ask yourself: *Which kind of impossible is it? Impossible of the first kind? Like 1 + 1 = 3?*

Or could it be the second kind of impossible, something very unlikely but absolutely worth pursuing if there were somehow a plausible way to make it happen?

I imagine that most would agree that my dream of finding Khatyrka's parent asteroid is far too wild an idea to take seriously. But if my three-decades-long search for a natural quasicrystal has taught me anything, it is this: Pay attention whenever someone says something is impossible and take the time to reach your own independent judgment.

Space science is moving in exciting directions. NASA is currently planning an Asteroid Redirect Mission (ARM) to visit a large near-Earth asteroid. Sometime in the 2020s, they hope to move the asteroid into a stable orbit around the moon and recover tons of materials from its surface for further study.

Most likely, Khatyrka's parent is in the asteroid belt and still orbiting the sun. Matthias Meier, our colleague who performed the crucial Zurich isotope experiment, has pointed out to me that carbonaceous chondrites, which include metallic aluminum-copper and aluminum-nickel alloys, might reflect sunlight differently than typical meteoritic minerals, at least for some wavelengths. That insight might help us to narrow the list of potential family members.

Suddenly, a completely far-fetched idea seems a bit less impossible. Testing has established that the Khatyrka meteorite snapped off from its parent asteroid two to four million years ago. Knowing the typical speed of an asteriod through space, it is possible to approximate where its parent might be in the asteroid belt. By studying the reflections of sunlight from the asteroids in that region, one might be able to identify an asteroid with the same chemical composition as our little Earth-bound orphan, Khatyrka. There are many uncertainties that could throw off such a calculation. Frankly, it is not even clear if it is a viable approach.

But Matthias and the rest of our team have already made a first attempt and found a possible parental candidate. It is an asteroid known as Julia 89 that lies in the main asteroid belt between Mars and Jupiter and orbits the sun about once every four years. About 150 kilometers across, Julia 89 belongs to a family of asteroids that formed during a collision several hundred millions of years ago, which is roughly the same time that Khatyrka is supposed have experienced a huge impact. It reflects light with a spectrum one would expect from a CV3 chondrite.

Now ask yourself: Could you picture an expedition landing someday on Julia 89 and discovering Khatyrka's secret?

Or would that be impossible?

ACKNOWLEDGMENTS

My scientific curiosity was sparked at an early age by my father, a masterful storyteller who often sat me on his knee and told me the most wonderful bedtime stories. My first memories date back to the age of three. Some nights, he would spin mythical tales about giants and dragons. The stories that truly mesmerized me, though, were the real-life ones about scientific struggles to unlock the secrets of nature.

I remember hearing about people like Marie Curie, Galileo Galilei, and Louis Pasteur. The scientists who starred in my bedtime stories were always more exciting than any imaginary dragon-slayer. The moment of discovery was always the climax—the instant when a scientist realizes a truth that no other human being has ever known before. My father would always dwell on that feeling, and never bothered to mention the fame that followed. The stories left an indelible impression on me. I desperately wanted to experience that same feeling. From then on, science was my passion.

I'll never know why my father chose to tell me stories about scientists and their grand adventures. He was a lawyer and, as far as I know, had no scientific training. He passed away from cancer when I was eight years old, long before the lasting influence his stories would have on my life became clear.

My general scientific attitude was strongly influenced by Richard Feynman while I was an undergraduate at Caltech. Other research advisors of mine there—Barry Barish, Frank Sciulli, and Thomas Lauritsen—and my PhD advisor at Harvard, Sidney Coleman, contributed greatly to my scientific development, especially in the areas of particle physics and cosmology. Others played an important role in setting me on the path to explore the structure of matter. Richard Alben, Denis Weaire, and Michael Thorpe were my mentors during a summer research program at Yale University in 1973. I worked for a dozen summers at the IBM Thomas J. Watson Research Center in Yorktown Heights, New York, with Praveen Chaudhari—a consummate scientist, mentor, and friend who encouraged me to develop my ideas about amorphous solids and, later, quasicrystals when few others would even consider them.

During my formative years at the University of Pennsylvania, I benefited from the support of my senior colleagues Gino Segrè, Ralph Amado, Tony Garito, Eli Burstein, Paul Chaikin, and Tom Lubensky. They encouraged me from the very beginning, despite the fact that the idea of quasicrystals seemed too fanciful to go anywhere. Tom patiently taught me the theoretical principles underlying condensed matter physics, and Paul introduced me to many creative experiments in his laboratory during the early years when quasicrystals first became known. They became my mentors, collaborators, and good friends. I also had the good fortune of having great students, including Dov Levine and Joshua Socolar, who made many key contributions.

When the search for natural quasicrystals began in earnest in 1998, a coterie of new people with extraordinary talents became part of my life, as described and named in this book, culminating in an unimaginably grand adventure followed by a cutting-edge scientific investigation that continues to this day.

In all of these endeavors, my role has been that of a conductor, a spectator, and always an admirer.

I cannot overemphasize the importance of our anonymous benefactor, Dave, who completely funded our scientific expedition to Chukotka. It is only because of Dave that our journey was possible and that there is a story to tell. Since the publication of the hardcover version of the book, I have received permission to reveal who "Dave" is: so now I can proudly and publicly thank David Bunning of Evanston, Illinois, for being such a magnificent friend of science.

Except for the expedition, essentially all of the research performed as part of the quest was done without any explicit grant support. My colleagues volunteered their energy, skills, and laboratory equipment for the benefit of science, drawing on discretionary funds and their own spare time. Everyone was eager to push the boundaries of science and to satisfy their own rapacious curiosity.

In addition to those explicitly mentioned in my story, there are many other people who contributed in different ways over the past four decades, a few of whom I would like to recognize and thank here. In exploring the fundamental physics of quasicrystals: graduate students Kevin Ingersent, Hyeong-Chai Jeong, and Mikael Rechtsman; senior scientists Marian Florescu, Paul Horn, Stellan Ostlund, S. (Joe) Poon, Sriram Ramaswamy, and Salvatore Torquato; photonics startup collaborators Joe Koepnick, Ruth Ann Mullen, Ben Shaw, and Chris Somogyi. In the science related to natural quasicrystals: students Ruth Aronoff and Jules Oppenheim; senior scholars John Beckett, Chris Ballhaus, Ahmed El Goresey, Russell Hemley, Jinping Hu, Mikhail Morozov, Jerry Poirer, Paul Robinson, George Rossman, and Paul Spry. In advice and support in recruiting financial support at Princeton: President Chris Eisgruber, Thomas Roddenberry, and James Yeh. In valuable advice on geology and expedition preparation: Wilfrid Bryan. In administrative and computer support before, during, and after the expedition: Charlene Borsack, Debbie Chapman, Laura Deevey, Vinod Gupta, Angela Q. Lewis, Martin Kicinski, and Alexander (Sasha) Tchekovskoy. My Russian tutor: David Freedel.

To be sure, the story of quasicrystals told here is a small part of a much larger international scientific endeavor. The account in this book is my personal perspective, rather than an objective third person history of the subject. There are many other creative scientists, mathematicians, and engineers around the globe who have made important contributions to the understanding of quasicrystals, many of whom are not named, including dear friends. It would not be practical or meaningful to list them all. But each of them has been instrumental in creating a new field of science and has my heartfelt gratitude and admiration.

A special source of inspiration for me has been my son Will. As his father, I cannot express the pride I feel having observed the intelligence, maturity, good humor, patience, and courage he exhibited during our time together in Chukotka. He had good reason to be concerned about me, but it never showed. Instead, he was a stalwart companion, advisor, fellow scientist, photographer, teacher, tireless worker, and loving son. Truly an inspiration.

This book would never have come to fruition without the contributions of an invaluable friend, Kathryn McEachern. I am immeasurably grateful that Kathryn volunteered her talents to help me tell this complex story through her tireless attention to detail, meticulous editing, and stubborn perfectionism, and all with good humor and boundless imagination.

I am thankful to the legendary writer and my Princeton colleague John McPhee, who shared his priceless advice about writing and story structure with me, and to Lincoln Hollister and Will Steinhardt for reviewing a draft of the manuscript. I am also thankful to my literary agents, John Brockman and Katinka Matson, who matched me with the marvelous team at Simon & Schuster including my editor, Jonathan Cox, who supported and shepherded me patiently through many rounds of revision with flexibility and wisdom. Many thanks to my cover designer Alison Forner, my production editor Kathryn Higuchi, my copy editor Frank Chase, my designer Ruth Lee-Mui, my legal counsel Felice Javit. and my publicist Elizabeth Gay. I am grateful to Sasha Kostin, Glenn MacPherson, Chris Andronicos, Chi Ma, Luca Bindi, Lincoln Hollister, Dov Levine, An-Pang Tsai, Peter Lu, Nan Yao, and my son Will for images, and to Rick Soden for photos of models and for preparing all the photo files for publication.

Last but not least, I thank my family, friends, and scientific collaborators for sharing their love, support, and brilliance with me. This book is a paean to those individuals.

IMAGE CREDITS

45 (*bottom*): Image by Edmund Harriss; Wooden Penrose tiles made by Edmund Harriss, Image and Tiles © Edmund Harriss

48: Image by author

50: Image by author

53: Image by dix! Digital Prepress

56: Image by dix! Digital Prepress

57: Image by author

58: Image by author

59 (*left*): Image by dix! Digital Prepress

59 (*right*): Image by dix! Digital Prepress

60: Image by author

66: Photo by Richard Soden

67: Image by author

69: Photo by author

71: Photo by author

74 (*left*): Photo courtesy of An-Pang Tsai

74 (*right*): Photo by author

75 (*left*): Image from Wikimedia Commons, by Vassil

75 (*right*): Image from Wikimedia Commons, by Materialscientist

77 (*top*): Image by dix! Digital Prepress

77 (*bottom*): Image by dix! Digital Prepress

83: Photo by author

90 (*top*): Photo by Richard Soden, *New York Times*, January 8, 1985

90 (*bottom*): Photo by Richard Soden, *New York Times*, July 30, 1985

92: Image by author

111: Photo courtesy of An-Pang Tsai

120 (*left*): Photo by author

120 (*right*): Image by Nan Yao

121: Images by author

122: From Bindi, Steinhardt, Yao, and Lu, *Science*, Vol. 324, 1306–1309, June 5, 2009

129: Photo by Glenn MacPherson

137: Photo courtesy of Nan Yao

141: Photo courtesy of Nan Yao

143: Photo by Nan Yao

144: Photo by Nan Yao

145 (*left*): Photo by Nan Yao

145 (*right*): Photo by Nan Yao

149 (*left*): Photo courtesy of An-Pang Tsai

149 (*right*): Photo by Luca Bindi

150: Photo courtesy of Lincoln Hollister

164: From L.V. Razin, N.S. Rudashevskij, N.V. Vyalsov, *Zapiski Vses. Mineral. Obshch.*, Vol. 114, 90 (1985).

166: Maps by Paul J. Pugliese

187: From Bindi, Steinhardt, Yao, and Lu, *Science*, Vol. 324, 1306–1309, June 5, 2009

195: Photo by Luca Bindi

199: Photo by Luca Bindi

203: From L.V. Razin, N.S. Rudashevskij, N.V. Vyalsov, *Zapiski Vses. Mineral. Obshch.*, Vol. 114, 90 (1985).

217: Photo by Luca Bindi

224: Image by dix! Digital Prepress

227: Image by author

231: Photo by William Steinhardt

250: Photo by William Steinhardt

252: Photo by William Steinhardt

257: Photo by Glenn MacPherson

260: Photo by William Steinhardt

261: Photo by Alexander (Sasha) Kostin

263: Photo by William Steinhardt and Richard Soden

269: Photo by Alexander (Sasha) Kostin

271: Photo by William Steinhardt

279: Photo by Glenn MacPherson

280: Photo by Alexander (Sasha) Kostin

282: Photo by Alexander (Sasha) Kostin

284: Photo by author

295: Photo by author

304: Photo by Luca Bindi

305: Photo by Luca Bindi

312: Photo by Glenn MacPherson

315: Photo by Luca Bindi

328: Photo courtesy of Chaney Lin

329: Photo by Nan Yao

332: Photo by author

340: Photo courtesy of Henner Busemann, ETH Zurich

344: Photo by Luca Bindi

345: Photos by Luca Bindi

347: Photo by Chi Ma

348: Photo by Chi Ma

350: Photo by Paul Asimow and Richard Soden

351: Photo by Paul Asimow

353: Photo by Chi Ma

356: Photo by Chi Ma

359: Photo by author

IMAGES IN COLOR INSERT

Image no.

1: Image by author

2: Photo by Richard Soden

3: Photo by Peter Lu

4: Images by Peter Lu and author

5: Photo by Luca Bindi

6: Photo by Luca Bindi

7: Image by Luca Bindi

8: Photo by William Steinhardt

9: Photo by William Steinhardt

10: Photo by author

11: Photo by William Steinhardt

12: Photo by William Steinhardt

13: Photo by author

14: Photo by William Steinhardt

15: Photo by William Steinhardt

16: Photo by William Steinhardt

17: Photo by Alexander (Sasha) Kostin

18: Photo by William Steinhardt

19: Photo by William Steinhardt

20: Photo by Alexander (Sasha) Kostin

21: Photo by Alexander (Sasha) Kostin

22: Photo by author

23: Photo by Alexander (Sasha) Kostin

24: Photo by William Steinhardt

INDEX

Page numbers in *italics* refer to illustrations.

ABOUT THE AUTHOR

PAUL J. STEINHARDT is the Albert Einstein Professor in Science at Princeton University and director of the Princeton Center for Theoretical Science. He has received the Dirac Medal and other prestigious awards for his groundbreaking theories of the early universe and novel forms of matter. He is the coauthor of *Endless Universe* with Neil Turok, which describes the two competing ideas in cosmology to which he contributed. In 2014, the International Mineralogical Association named a new mineral "steinhardtite" in his honor. A fierce defender of science and scientific reasoning, Steinhardt continues to challenge conventional thinking and identify new directions ripe for exploration and innovation.